What's
in Your
Genes?

Published by
Adams Media, a division of F+W Media, Inc.
57 Littlefield Street, Avon, MA 02322. U.S.A.
www.adamsmedia.com

ISBN 10: 1-4405-6764-6
ISBN 13: 978-1-4405-6764-3
eISBN 10: 1-4405-6765-4
eISBN 13: 978-1-4405-6765-0

Printed in the United States of America.

10 9 8 7 6 5 4 3 2 1

Library of Congress Cataloging-in-Publication Data

McKissick, Katie.
What's in your genes? / Katie McKissick, creator of Beatrice the Biologist.
   pages cm
Includes bibliographical references and index.
ISBN-13: 978-1-4405-6764-3 (pbk. : alk. paper)
ISBN-10: 1-4405-6764-6 (pbk. : alk. paper)
ISBN-13: 978-1-4405-6765-0 (ebook)
ISBN-10: 1-4405-6765-4 (ebook)
1. Genetics. I. Title.
QH437.5.M384 2014
572.8--dc23

2013031102

Illustrations by Katie McKissick.
Cover images © Katie McKissick.

This book is available at quantity discounts for bulk purchases.
For information, please call 1-800-289-0963.

What's in Your Genes?

FROM THE COLOR OF YOUR EYES TO THE LENGTH OF YOUR LIFE, A REVEALING LOOK AT

YOUR GENETIC TRAITS

KATIE McKISSICK
Creator of Beatrice the Biologist

Avon, Massachusetts

Dedication

For my grandfather, Dr. Michael Lyons to some, who to me was and will always be Papa.

For always encouraging me and pointing me in the direction of science.

For indulging my curiosities, taking me to the medical school to play with high-powered microscopes and see parts of cadavers thawing in the sink.

For suggesting that I be a science writer before I could fathom that I would ever even be capable of such a thing.

When I visited him in the hospital, I typed at his bedside while he slept. I'm glad I was able to tell him that my science writing dreams were coming true. When he woke up from a nap, I told him I was working on a book about genetics. He said to me, "I can't wait to read it."

I love you, Papa. Rest in peace.

Acknowledgments

I would like to thank my husband Kip Barnes, my mother Maureen McKissick, and my godmother Anne Bruce for their unending support and guidance during this whirlwind writing process. Instead of the usual "I couldn't have done it without you," I'll say that without you in my life, I would be in the fetal position in the corner of a house I was squatting in, rocking back and forth muttering to myself. So obviously, writing a book would have been out of the question from the start. Thank you a million times over and then some.

Contents

PART 4: BEYOND YOUR GENES 201

Introduction

Most of us would like to think we are special and amazing because of our essential, mysterious self-i-ness, but it's really because of our DNA. Experiences of course matter too, but much of who we are and who we become is written inside of us. While it's romantic to think that our identities can't be quantified, I think it's more poetic that we have it inscribed in every part of our bodies instead.

But don't fret—you still have that wonderful you-ness that no one else has, for no one else has your DNA sequence (at least, statistically speaking I'm pretty comfortable saying that, provided that you don't have an identical twin), and certainly no one has your unique past of influential experiences.

The beauty of peering inside our DNA is that it contains all the stories of our past, linking us to our deepest of roots—our parents, grandparents, great-grandparents, all the way back to our ancient ancestors, some of whom were human, others who were not. It's a book that's been revised, rewritten, and edited over the course of more than three billion years. Through it, we are connected to every other living thing on the entire planet that is alive today or has ever lived. This is serious stuff, man.

Teasing the secrets from our DNA is a work in progress, so you need not worry that someone can steal a piece of hair or some chewed gum of yours and with scientific finagling see into your soul. What we know and don't know about genetics is constantly changing. Many of the things that I say here are still under investigation, and the details will be ironed out in just the next few years—it's all moving that fast. Some of the details that scientists are relatively certain of today could be proven wrong tomorrow. It's a pretty tumultuous area, genetics. But what won't ever be out of date are my musings about its importance and its intrigue. I'm always in style (at least I hope so).

Genetics has so many terms it's more like speaking another language than just learning about heredity. Science is known for its jargon, certainly, but genetics is particularly vocabulary heavy. That's why there's a glossary at the end of this book for easy reference. Despite the plethora of terms, I will at all costs avoid turning this book into a steaming pile of vocabulary lessons. I will use the authentic science terms when I feel it's valuable, but I'll gloss over them or use synonyms when I feel that it's not. This may anger some of my scientist brethren, but to them I say, "Pffft. Whatevah."

In the pages that follow, I will draw rudimentary cave paintings of chromosomes for you, make dirty jokes about proteins, and make fun of just about every historical scientist possible.

If you are looking for a serious take on the science of heredity, you won't find your fix within these pages. There are plenty of textbooks out there for that (they are also effective doorstops). What I want to do is cover all the bases of genetics while entertaining you. I'll distill the most cutting-edge research (which surely will be outdated in just a few years) and historical context into a conversation that I would gladly have over coffee with anyone willing to listen, and will have with you via the following pages.

Science is serious business, to be sure, but it doesn't always need to be taken so damn seriously. While it may be our species' greatest achievement, responsible for every technological advance we have or ever will make, science is also poop and sex and boogers. So let's put our feet up and discuss DNA drama, gardening-crazed monks, and superpowers. Because that is what genetics is really all about.

PART 1
The Basics

Your Friend, DNA

DNA isn't much of a showboat. It stays hidden away inside your cells, preferring to be left alone, keeping all its secrets. (Sounds a lot like my ideal vacation.) But sorry, DNA. We're going to be intruding on that personal space of yours, whether you like it or not.

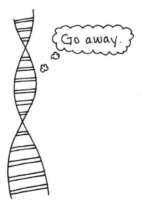

DNA is a useful acronym, because the full name is a bit of a mouthful: deoxyribonucleic acid. Yeah, I know. It's not a great name, but don't blame me. I wasn't around when it was named. I would have saved everyone a lot of time by naming it Reginald, or maybe Gladys. But this is the name we're stuck with. I know it looks daunting, but when you split it apart, it's not so bad.

DNA: An acronym for deoxyribonucleic acid, which is really boring stuff that is found in all of your cells and contains all the information for making you who you are. You know. No big deal.

Deoxy is just there to say that something has one less oxygen atom. What exactly has one less oxygen? The next part of the name, *ribo*. That's short for a sugar called *ribose*. So the whole first part, "de-oxy-ribo" just means that there is something in it called a ribose that's missing one oxygen atom. So far, so good.

Deoxyribose: A pentagon-shaped sugar that is found in the backbone of DNA.

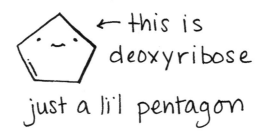

Now let's get to that ribose bit I mentioned. Ribose and deoxyribose are sugars. "-ose" is actually a very popular ending for various types of sugars. (Think of glucose, sucrose, and fructose.) Deoxyribose isn't nearly as popular as these other sugars, but it should be, since it's an important part of DNA. Deoxyribose is part of the backbone of DNA—the sides of the twisted ladder.

The other part of the backbone is something called a phosphate group. It's made of an atom of phosphorous and four atoms of oxygen. The sides of the DNA ladder are alternating phosphate groups and deoxyriboses. My word processor doesn't think that deoxyribose is a real word, but I'll have to assure it that it is.

Phosphate: A chemical compound that makes up part of the backbone of DNA. It is a phosphorous atom surrounded by four oxygen atoms.

a phosphate group

But the backbone of sugar and phosphates is not exactly the fun part of DNA. The real business-doing portion of DNA is in the rungs of the ladder where the "bases" live. You may have heard of the bases of DNA before—they're those letters A, T, C, and G. Their full names are adenine, thymine, cytosine, and guanine. It's pairs of these guys that make the rungs of that DNA ladder. The bases are very particular about how they go about this: A pairs with T, C pairs with G. I heard that A can't stand to be around C, and G thinks T is an obnoxious brat. But you didn't hear that from me.

Base: The part of DNA that forms the rungs of the double helix ladder.

To describe the base pairs matching up just so, we say that they complement each other. If you read down one side of DNA and have TTAAGC, the complementary sequence would be AATTCG. Each base has a specific complement, which is not to be confused with bases complimenting each other, which I'm sure they often do, but it has not yet been scientifically verified and published in a reputable journal. I'm still holding out hope, though.

Adenine: One of the bases in DNA that make up the rungs of the ladder. It pairs with thymine.

Cytosine: One of the bases in DNA that make up the rungs of the ladder. It pairs with guanine.

Guanine: One of the bases in DNA that make up the rungs of the ladder. It pairs with cytosine.

Thymine: One of the bases in DNA that make up the rungs of the ladder. It pairs with adenine.

To sum up: DNA is made of equal parts deoxyribose (that sassy sugar), phosphate groups (just a group of five atoms), and a base (A, T, C, or G). Together this unit is called a nucleotide. You'll notice *nuc* in lots of names. It seems scientists were smitten with the word *nucleus*, and since all of these things spend lots and lots of time in the nucleus of the cell, they just couldn't help but include it in the name.

BEHOLD! A nucleotide:

phosphate

sugar

base

Nucleotide: The basic building block of DNA. It is made of a sugar, a phosphate, and a base.

When we talk about a sequence of DNA, we're not reading the base *pairs*, we're just talking about the order of the letters as you go along one side of the ladder. Here's an example of a DNA sequence:

ATGCCGCGCGTTTCGATATCGCTTTTCGCGAAAAAA AAA

Yup, that's what it all looks like. Pretty exciting, eh? Those four letters in random orders just going on and on like that. Your DNA is a very rambly book, about 3 billion letters long. If this book had 3 billion letters in it, it would be more than 60 million pages. And if that were the case, I wouldn't ever write it, as I would die from typing exhaustion. And old age. I'd also probably run out of things to say and just start smashing my forehead against my keyboard, and even if I did that for years I'd never finish. So ... yeah, it's really long.

And remember, that 3-billion-letter instruction booklet of yours has just four letter varieties. Just take that in for a moment. Every bit of information in your DNA that makes you who you are—your tastes, your eye color, your obsession with '80s romantic comedies—is expressed and stored with just four measly little letters.

In English, I'd run out of ATCG word combinations real fast. *At.* There's one. *Cat.* On a roll here. *Tag.* Yeah, okay. I'm done with this. You win this one, DNA. But don't let it go to your head.

DNA, the Copycat

The twisted ladder shape of DNA, and the fact that the rungs are matched pairs of bases, isn't just for looks. This structure makes DNA very easy to copy, just like history homework. It's actually quite amazing. If someone took a page out of this book (and I sincerely hope no one feels the urge to do that), ripped it in half (again, please don't do this), and gave it to you, you wouldn't be able to get much out of it. With half the words missing, it would be rather useless as reading material. It might still be useful for paper airplanes, but not much else. But if DNA is ripped in half down the rungs of the ladders (which does happen), you can tell exactly what the other half of the strand would look like. This is a really handy feature.

One of the ways DNA actually makes use of this wonderful attribute is in DNA replication—when DNA makes more of itself. And replication is something DNA has to do frequently. Anytime your body has to make a new cell, it needs more DNA for that little baby cell.

Replication: The process of DNA making a copy of itself.

Your body is making new cells all the time. For example, think about your skin. You have a completely new coating of skin cells on your body every six weeks because you are constantly making more of them. And that means a lot of DNA replication is going on there.

In truth, the skin cells that you see when you look at your hand right now are all dead. The new skin cells are coming from below to replace the dead cells you're constantly shedding as you go through your day. Where do those dead cells go? See that bit of dust on your elliptical machine? That's where they went.

Dust: it's not just dirt and fly eggs, it's also your dead cells! Yummy!

DNA has to make an entire copy of itself before every cell divides so that all your new cells have a full set of "instructions." How DNA actually gets copied is a complex ballet of chemistry and chaos.

The DNA first has to be unzipped—which leads to many semi-clever jokes about genes (read as *jeans*) being "unzipped." Har har.

When it gets split down the middle, the bonds between the bases are broken, so you have the two sides of the DNA hanging out in the open, fully exposed like a molecular skinny dipper. To get two new full strands of DNA out of this unzipped, floppy one, nucleotides that are just floating around swoop in to be paired up correctly with the lonely bases there. Adenines are connected to thymines; cytosines are paired with guanines.

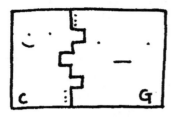 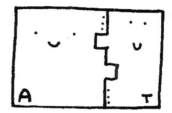

In biology class, these floaters are called "free nucleotides." They're "free" because they're not attached to anything, but they also cost zero dollars, so they are in many ways free nucleotides. I bet they don't wear underwear or bras either.

Now that all the new nucleotides have been matched up, you have two complete strands of DNA, where there used to just be one.

Makin' Mo' DNA

T A		T	A
C G		C	G
A T	⇨	A	T
G C		G	C
G C		G	C

⇩

Look!
Now there's
two of 'em!

T A		T A
C G		C G
A T		A T
G C		G C
G C		G C

During replication, you can also see the other bonus of DNA's strand-y, ladder-y structure. These two new strands are made of half of an old strand and half of a new strand. This is clever because it provides insurance against mistakes, as using the original as a template makes it far more likely that you matched up the bases correctly. It doesn't always work out, though. Even with 99.99999 percent accuracy, when there are so many bases in your entire set of DNA, you are going to wind up with some mismatches and mistakes. That's one reason that your body has proteins that scan the DNA looking for typos to fix. But with 3 billion letters to look at (actually, double that if you think about both sides of the DNA), there are bound to be some oopsies.

In addition, your DNA can only be copied so many times before it starts to look a little old and tired. Just like making a photocopy of a photocopy of a photocopy, the quality starts to degrade. Some of a body's changes during the aging process are due to mistakes in DNA replication and the shortening of the DNA strands. Every time DNA is copied, a tiny bit on the end gets lopped off. To protect against this, DNA strands have protective caps on them called telomeres. But after years and years of replicating, those buffers are slowly eaten away, and then you're out of luck.

If our lifespan is partly determined by our DNA getting "old" as it gets copied over and over, what about species that live seemingly forever?

Have you heard of the "immortal jellyfish"? (First of all, before you get angry about fishes of jelly, I know it's not a fish, and that the better term is just *jelly*, but *immortal jelly* sounds like a menu item at a New Age restaurant, or a euphemism for a sex act, so I used my editorial powers and went with "jellyfish." It's also just more satisfying to say.)

These immortal jellies of fish of the sea appear to be immortal, and not simply because they haven't been observed dying of old age.

After they mature, they start to age in reverse, reverting back to the baby form of a jellyfish, which is a polyp.

Polyps are stationary little blobs that rest on the sea floor and look a little bit like anemones. The jellyfish you picture as the bell with dangling tentacles, gracefully floating in the ocean, is called the medusa stage of their life cycle.

These immortal jellies go forward and backward developmentally, going from baby to adults, back to baby, and repeat. It's like Benjamin Button, except it keeps cycling through.

IMMORTAL!

How are they doing this? How does their DNA not show signs of wear and tear as these life cycles are played out again and again and again? We're working on figuring that out. And by "we" I mean the royal science "we," as in people who have nothing whatsoever to do with me.

Now that you know what DNA is and how it wound up in every one of your cells, we need to get to the more important stuff, like what the hell it does.

Team Protein

DNA happily takes all the credit for all this *making you who you are* business, but the truth is that DNA doesn't really do much of anything. DNA by itself is like a tree falling in the woods with no one to hear it—that is, a nonevent. DNA is really just passive and lazy. It only has so much sway because it gets help from RNA and protein.

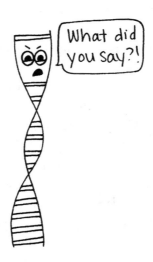

Oh, RNA, another one of those annoying acronyms. But this time, there is no D for deoxy. That part is gone from the name, so RNA is just *ribo*nucleic acid. This time the sugar is ribose instead of deoxyribose. The difference is an oxygen atom—ribose has one more oxygen atom than deoxyribose does. Let's just say that they are very, very similar in their basic structure, save for one measly oxygen atom.

RNA: An acronym for ribonucleic acid, which is usually a single-stranded string of bases that is made from a sequence of DNA. Some RNAs do jobs in the cell, and others are used to make proteins.

Protein: A chain of amino acids that twists and folds into an absolutely insane shape. They do just about everything of importance in your body.

Ribose: A pentagon-shaped sugar that is found in the backbone of RNA.

The rest of the name, nucleic acid, was just what some guy named it before we really knew what it was. Nucleic got tacked on there because it's found in the nucleus, and acid because, well, it's slightly acidic. This is the irony of scientific names: it's only in an attempt to make things simple and clear that they become so overwhelming and confusing.

There are a few different kinds of RNA, but I'm just going to talk about messenger RNA, which is acronym-ed as mRNA. (Are you sick of acronyms yet? I am!)

Messenger RNA is single-stranded, which means it's not a ladder like DNA; it's just a floppy noodle. mRNA is probably jealous of DNA's lovely helical ladder business at times, but mRNA leads a much more romantic life because, as the name suggests, it takes information and delivers it somewhere else. It's on the move and seeing the world. But DNA gets to live vicariously through it.

Atlanta-Fulton Public Library System

Metropolitan Branch
www.afpls.org

Checked Out Items 7/27/2018 12:20
XXXXXX5608

Item Title	Due Date
R4001289940	8/24/2018

What's in your genes? : from the color of your eyes to the length of your life, a revealing look at your genetic traits

1332 Metropolitan Parkway
Atlanta, GA 30310

Have a great day!

This is really the importance of RNA. It gets stuff done. Like I said before, DNA is just a couch potato (or nucleus potato, let's say). Like a paranoid, antisocial conspiracy theorist, it never leaves the house. But it has all this vital information that needs to get out, so it sends mRNA to do its bidding. And the important business that mRNA tends to is helping to make proteins.

Proteins are what it's all actually about. Yes, DNA is the instruction manual for building a person, or a frog, or a dinosaur. But what those instructions are for, specifically, is making proteins. Lots and lots of different proteins.

We usually only talk about proteins in a dietary way, such as when people ask vegetarians, "How do you get enough protein?" Or when weight-lifting gym rats crave a protein shake after a workout. Or when commercials boast that a nasty fiber bar has five grams of protein. (I don't care. I'm not eating that brick.)

But protein is far more than food. It's really everything about you. Hormones such as testosterone and estrogen, which determine whether you're male or female, are proteins. The hair on your head and the nails on your fingers (and the other places you have hair and nails) are proteins. The parts of your cells that allow you to move your muscles are made of proteins. You, my friend, are a great big gorgeous bucket of proteins. And those proteins were made with the help of mRNA, which got its instructions from DNA.

Proteins come in a lot of different shapes and sizes, but they are all made of chains of things called amino acids. There are 20 amino acids, and they have really fun names like lysine, tyrosine, and asparagine. If you're an avid movie quoter like myself, you may remember that in *Jurassic Park*, one of their containment techniques was called the "lysine contingency." They engineered the dinosaurs to be incapable of making lysine, so that if they were to escape the island and be without special *Jurassic Park*–provided lysine-supplemented food, they would die. I can still hear Samuel L. Jackson's voice explaining it. Those guys thought of everything!

Except fat programmers causing power failures. They didn't plan for those.

Amino acid: The building blocks of proteins.

It just so happens that we, too, are incapable of producing lysine. There are nine amino acids that we refer to as the "essential amino acids," specifically because our bodies do not make them. *Essentially* (tee hee), we have to get these in our diet, just like the cloned *Jurassic Park* dinosaurs. They are:

1. histidine
2. isoleucine
3. leucine
4. lysine (just like the *T. rex*!)
5. methionine
6. phenylalanine
7. threonine
8. tryptophan
9. valine

You don't need to worry about the dietary amino acids much, because as long as you're eating more than Altoids and water, I'm sure you get all of them every day. Animal proteins (meat, milk, eggs, and cheese) have all these essential amino acids in them, so we call them "complete proteins."

But our vegetarian friends don't need to worry too much about getting all these amino acids, either. Although some plant-based foods may lack an amino acid here or there, as long as your diet is varied, you will cover all your bases. For instance, corn is not high in lysine or threonine, but beans are. To the taco truck!

A really small protein would be fewer than 100 amino acids in length, and the biggest proteins are thousands of amino acids long. But no matter how big or small a protein is, it's all but guaranteed to look like a hot mess. I mentioned that a protein is a chain of amino acids, but it doesn't actually look like a chain—unless your chain is twisted and looped and clumped into a ball, in which case you have probably ruined a perfectly good chain. Proteins are actually folded into a three-dimensional shape when they are created.

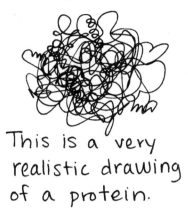

This is a very realistic drawing of a protein.

A protein's shape, however messy it appears, is vital to its purpose. Proteins interact with a lock-and-key-type arrangement. If one of your locks or keys is misshapen, any number of things could happen, from as banal as blue eyes all the way up to a whole list of genetic disorders.

Protein folding is so complicated that some scientists spend their entire careers figuring out the exact three-dimensional shape of specific proteins. With the twists, turns, bends, loop de loops, corkscrews, and other such insanity, proteins are a big ol' complicated train wreck. But they do very important work, so we'll have to give them a break.

The Genetic Code

DNA may provide instructions on how to make those important proteins, but actually making one of those proteins involves a lot of players. First, the DNA needs to make a temporary copy of itself that can leave the nucleus. DNA doesn't leave the safe haven of the nucleus during the normal hustle-and-bustle of the cell's business (it does when the cell divides, though). It's too dangerous out there, and DNA is just too important to deal with the peasants outside of its castle. Instead, DNA unwinds and has a copy of itself made as a single strand of messenger RNA.

RNA is a lot like DNA, but has a few key differences. Like I said before, the sugar in the backbone is ribose instead of deoxyribose, which gives it a different acronym. It's also a single strand rather than a ladder like DNA. And it has a different base, called uracil, that takes the place of thymine. When the DNA unwinds and exposes itself (and boy, is that awkward), a bunch of helper proteins use the bare DNA sequence to line up RNA building blocks that complement the sequence.

Uracil: A base (like adenine, thymine, cytosine, and guanine) that is only found in RNA, not in DNA. It takes the place of thymines in RNA, so it pairs with adenines.

If the DNA has a C, an RNA nucleotide for G will pair up with it. And uracil pairs with the A's in the DNA sequence. When the mRNA has been built to match up to the DNA sequence, it breaks free, and the DNA zips itself back up like a guy at a urinal—quickly, yet carefully. We now have the DNA back to normal and a new bit of mRNA that contains a mirror image of the DNA sequence we need.

I should also mention that part of the process of making mRNA involves removing filler sequences called introns. Our genes are a little odd in that they contain these unused sections that have to be deleted from the mRNA. If a gene was a book, it would be like inserting a blank page here and there, which the publisher would have to remove before printing. Introns are pretty weird. Evolutionarily, they insert some variety into the system. When the introns are removed, different parts and amounts could be removed, which is called alternative splicing, and you can essentially have two genes in one. It's strange.

Gene: A section of DNA that contains instructions for making RNA. That RNA may even be used to make a specific protein.

Once complete, the mRNA and DNA share a tearful goodbye, and the mRNA leaves the nucleus to fulfill its destiny.

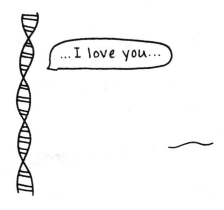

Out in the real world—the main body of the cell outside the nucleus—the mRNA looks for a ribosome. Ribosomes are very small cellular machines that connect with mRNA to actually build a protein. It does this by "reading" the sequence of the mRNA according to the genetic code.

Translation: The process of a ribosome reading a sequence of RNA and using it to build a protein.

The genetic code sounds really intense and scientific, but relax. It's actually quite simple (if not a little dull). The genetic code is a set of rules for how mRNA is translated into a protein. Really, that's all. It's just a list. When the ribosome goes along the length of the mRNA, it reads it in 3-base chunks called codons, like AGG, ACG, AUG. Each codon stands for an amino acid (the building blocks of protein), except for a few that are called "start codons" and "stop codons." They are exactly what they sound like. A start codon signals to the ribosome, "Start over here, brainiac." And when a ribosome gets to a stop codon like UAA, UAG, or UGA, it stops reading, and the protein is done. And there is much rejoicing.

Codon: A set of three bases in RNA that translates to one amino acid when proteins are being made.

Stop codon: A 3-base "word" in the RNA sequence that tells a ribosome to stop building the protein.

This is what the genetic code actually looks like:

The Genetic Code	
Letters in mRNA	**Amino Acid**
GCU, GCC, GCA, GCG	Alanine
CGU, CGC, CGA, CGG, AGA, AGG	Arginine
AAU, AAC	Asparagine
GAU, GAC	Aspartic Acid
UGU, UGC	Cysteine
GAA, GAG	Glutamic Acid
CAA, CAG	Glutamine
GGU, GGC, GGA, GGG	Glycine
CAU, CAC	Histidine
AUU, AUC, AUA	Isoleucine
UUA, UUG, CUU, CUC, CUA, CUG	Leucine
AAA, AAG	Lysine
AUG	Methionine
UUU, UUC	Phenylalanine
CCU, CCC, CCA, CCG	Proline
UCU, UCC, UCA, UCG, AGU, AGC	Serine
ACU, ACC, ACA, ACG	Threonine
UGG	Tryptophan
UAU, UAC	Tyrosine

The Genetic Code	
Letters in mRNA	**Amino Acid**
GUU, GUC, GUA, GUG	Valine
AUG	START
UAA, UGA, UAG	OMG, STOP!

The genetic code is a lot like DNA: all by itself it's actually pretty boring, and yet it is responsible for everything we see and do, quite literally: our eyes wouldn't work if we didn't have the proteins to work the light receptors, and we wouldn't be able to do or think anything if we didn't have proteins to get our brains and muscles working.

The genetic code isn't specific to us humans. All living things use it. It's probably the only thing we can all agree on, actually. Whether you have a nucleus or not, are a single-celled blob or a multicellular heap, whether you breathe oxygen or are poisoned by it, you use the genetic code. It's been around since the first cell fwooshed into existence 3.8ish billion years ago.

So now that you know the meaning of life—the genetic code—do you feel more fulfilled and worldly? No? Try ice cream then. That usually works for me.

Just Genes

AATCGCTAGCGATACGATCGTACGTAGCTAGCGCGCGA
CGCACGACGACGACGACACGACGGACGACGTACGTAGC
TAGCTGACTGATCGACGACGACTAGCATCGAGCAGCTG
ATGCGCGATGCTACGTAGCTAGTCACGACGACTACACG
ACGACTACGACGCATACG

Look, a gene! Well, a hypothetical, very short gene, that is. I just held my fingers over A, T, C, and G and went to town. I don't think this particular gene would actually code for anything useful.

But seriously, at this point, we probably ought to talk about what a single gene actually is. A gene is a portion of DNA that your cells make RNA from.

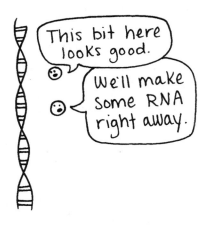

Another way of saying this is that a gene is a stretch of DNA that has instructions for making a specific RNA and/or protein, or that it "codes for" them. I was going to just say that a gene has directions for making a single protein and leave out the RNA part, but that wouldn't be quite correct.

Some DNA codes for RNA that doesn't end up getting used to make a protein. A good example of this happens to be the structure that helps make protein, the ribosome. In a bit of a strange twist, ribosomes are made of RNA. So you have RNA reading RNA to make protein. RNA is quite the busy bee.

True story:

Ribosomes look
like dog toys.

We humans have twenty-something thousand genes (we don't have an exact number just yet). But before we started actually counting them, most scientists thought we'd have over 100,000 genes (you know, because we're so very awesome). Alas, we do not have that many. We are not as cool as we thought. And quite a bit of our DNA doesn't actually code for anything at all. It just lolls about, relaxing. I'll talk more about this later.

Genes are named according to their function, or the disorder that results if the gene isn't working, a la the "breast cancer gene." There are actually several genes that live under the breast cancer gene umbrella. One is called, quite imaginatively, *breast cancer 1, early onset.*

Genes, just like elements on the periodic table, have names and a symbol. Not a cool hieroglyphic, visual symbol, but a shortened letter version, usually more or less an acronym. That same breast cancer 1 gene's symbol is BRCA1. BR for breast + CA for cancer + 1 for type 1. As if we needed any more acronyms! But numerical names would be worse.

"Hey, Olga, do you have a sample of the isolated 19,345 gene? I need it over at my bench."

"I thought you wanted 19,354. Dammit!"

Your full set of genes is your genome—your whole instructional guide. All your cells have a full copy, although they don't read the whole thing. The cells in your liver just read the liver cell chapters, muscle cells read muscle chapters, and pancreas cells read the pancreas chapters. Ever since you were just one tiny little cell you've been handing out full copies of your genome to every new cell you make. It's rather amazing that it has all worked out so well.

When I was looking into the gene-naming conventions currently used, I was surprised that the location of the gene on a specific chromosome didn't factor into the name. I was expecting the name to in part denote where it is in the genome. But this isn't up to me.

Do you know who this *is* up to? It's up to Hugo. Actually, it's HUGO, the Human Genome Organisation (yes, it's spelled that way—not everyone is from 'murica). And the names are up to the HUGO Gene Nomenclature Committee, the HGNC, whose members are scientists from sixty-nine countries that specialize in human genetics. I bet people think they sell vitamins.

Figuring It Out

When we think about big DNA discoveries, most people think about Watson and Crick, who in 1953 published a paper that showed their model for the structure of DNA. This *was* big-time news. But what I find surprising is that just one year before, in 1952 (thanks, math!), someone was doing experiments to show that DNA was the genetic material that makes each living thing what it is. You'd think that if Watson and Crick spent a bunch of time trying to nail down DNA's structure, we'd have some clue as to what DNA was for and what it does in the cell.

Nope. Right up to then, most scientists thought that proteins, not DNA, were the genetic material that is passed down through generations. They figured that the amount of information that would be required to have all the instructions to build a living thing couldn't possibly be stored in something like DNA—that proteins, with their highly complex and insane structures, simply had to be the manual for such a complicated IKEA product like ourselves. They thought DNA was too simple for the job. Oh, poor misunderstood DNA.

So how did we finally figure out that DNA was actually useful? With bacteria and viruses. Obviously.

Viruses are especially talented at infecting things. They attach themselves to cells, get inside, and take over. Those jerks. Back then, scientists knew that viruses were changing the cells somehow, but they didn't know what they were using to do it—was it DNA or was it protein that viruses injected into cells to hijack them?

Having to figure out what is happening on a microscopic scale is tricky when you don't have super intense real-time microscopes, and those were somewhat lacking in 1952. It was practically the dark ages; they didn't even have iPhones.

Virus: A piece of nucleic acid (sometimes DNA and sometimes RNA) surrounded by a protein shell.

So how do you find out what the hell is going on in the tiny, effectively invisible world of bacteria, viruses, protein, and DNA? You tag it with radioactive markers.

Alfred Hershey and Martha Chase grew bacteria-infecting viruses, which are called phages (rhymes with *pages*), in baths of radioactive elements. One batch was grown in a pool of radioactive sulfur, and another was grown with radioactive phosphorous. Why sulfur and phosphorous? Well, the sulfur will get incorporated into the proteins the phages make, and the phosphorous will get incorporated into the DNA. So what we now have are two groups of viruses, one with radioactive protein and the other with radioactive DNA.

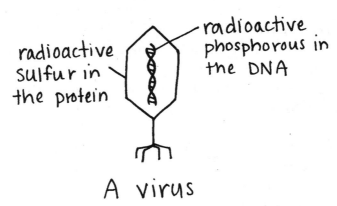

radioactive sulfur in the protein

radioactive phosphorous in the DNA

A virus

Now what did they do with this strange, glowing virus soup they made? They gave it a whole bunch of bacteria to infect. The poor bacteria didn't stand a chance. They were just inundated with an avalanche of viruses coming right at 'em. After Hershey and Chase gave the viruses plenty of time to infect as many bacteria as possible, they put the two different mixtures into blenders to separate the viruses from the bacteria.

Next, to condense the bacteria into one area to see if it contained radioactive DNA or radioactive protein from the viruses, they spun it in a centrifuge, the most violent merry-go-round in the world. It spins test tubes at tens of thousand of rotations per minute (notice I didn't say RPM because *down with acronyms!*), so the heaviest parts of the mixture sink to the bottom and form a glob. Hershey and Chase inspected the bacteria globs at the bottom of the test tubes to see if they had radioactive protein or radioactive DNA in there.

What they found was that the batch with radioactive protein didn't show any radioactivity in the bacteria—all of it was still with the viruses. In the radioactive DNA batch, all the radioactivity was concentrated in the bacteria. This meant that the viruses were injecting DNA—not protein—into the bacteria. This experiment proved that DNA, not protein, was the molecule that had the power of storing information—hereditary information for us, and the information that viruses required to infect cells. It's all because of DNA. This whole situation is now called the Hershey-Chase experiment, which would also be a pretty good band name, or some kind of terrible chocolate concoction.

The discovery of what DNA actually looks like followed just one year later. A lot of people were working on this mystery. It got serious when a fellow named James Watson was visiting his buddy Maurice Wilkins in London at King's College. There, he got a glimpse of a picture taken by Rosalind Franklin using something called X-ray crystallography. She was basically taking pictures of DNA using X-rays instead of regular light that normal photography requires. Because she spent so much time working around the radiation from the X-rays, Franklin died of ovarian cancer at the age of 38. In the scientific community, we call this a "Debbie Downer."

Watson knew a lot about X-ray crystallography, so when he saw that picture, it gave him all the information he needed to solve the puzzle of DNA's design. He and Francis Crick worked out the particulars, built a model of DNA, and published their results.

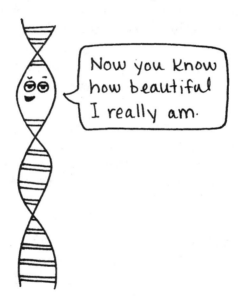

In 1962, James Watson, Francis Crick, and Maurice Wilkins shared a Nobel Prize for the discovery of DNA's structure. Rosalind Franklin was not part of the awarded team because, as a rule, Nobels are not awarded posthumously. But she was only dead because of the research she did that was so vital to Watson and Crick's work. It's just so very frustrating.

I hereby award Rosalind Franklin the McKissick Award for Scientific Excellence and Damn Awesome Crystallography. I only hope this small gesture can right some of the wrongs in this strange story of scientific discovery. Rosalind, you the woman. (This is where we'd fist-bump in my imagination.)

You're a Mutant

We're all mutants. I know the word *mutant* has a lot of baggage, though: either the good kind like X-Men and other such mutation-caused superpowers or the negative connotation that suggests misplaced eye sockets or extra appendages. But anyone with even a small mutation could be considered a mutant. Blue eyes are nothing more than a mutation in the gene for eye pigmentation. That makes me a mutant for sure.

A mutation is any old change in DNA. It could be as little as a single base pair's worth of difference. But even then, there are a few different varieties. A single base pair could be completely removed from the sequence, could be changed to a different letter, or could be wedged in there erroneously. If something is different, say a T is changed to a G, that is probably not that big of a deal, but if a base is completely removed or one is inserted, that can cause serious damage.

Mutation: Any change in DNA. It could be very small, just one letter, or it could be a missing or extra segment.

Why would adding or subtracting just one letter make such a big difference? Well, when the DNA sequence is used to make a strand of mRNA, and that mRNA is used to make protein, it gets read in those three-base segments called codons. If you add a letter in there, you screw up every single codon after it. It's called a frame shift. Watch.

Let's say this is your DNA sequence:

AATTGGCCCGGAACT

This then would be the mRNA sequence that complements it:

UUAACCGGGCCUUGA

When that mRNA gets out into the cell, it meets up with a ribosome to make a protein. The ribosome reads it in three-letter chunks, so it sees this:

UUA ACC GGG CCU UGA

Leucine, Threonine, Glycine, Proline, Stop.

Let's see how completely different it looks if we add a letter in the fourth spot of our DNA.

AATC TGGCCCGGAACT

The mRNA isn't wildly different.

UUAG ACCGGGCCUUGA

But when we look at the codons, we have a serious problem.

UUA GAC CGG GCC UUG

Leucine, Aspartic acid, Arginine, Alanine, Leucine

Look! Everything after the first amino acid has been changed, so now we have a completely different protein on our hands. What have you done?!

The best way to completely ruin a gene is to change a regular codon into a stop codon, and it does happen. Imagine you're making a big protein of about 1,000 amino acids. The ribosome is moving along, reading codon after codon, adding amino acid after amino acid. But then it comes to the 501st codon, which due to a mutation has been changed from UGG, which codes for tryptophan, into UGA, which is a "stop codon." The ribosome has no way of knowing this is a mistake. It comes across a stop codon, and its job is done as far as it's concerned. It clocks out and heads home. But we are stuck with a protein that's half the length it should be. So you heard it here first. A single letter difference can cause a horribly mutated protein. Sometimes it's really that simple.

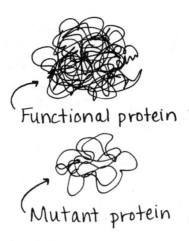

Functional protein

Mutant protein

And these are just the mutations that can happen from a single base's worth of difference. There are times that a section of DNA can move from one spot in the genome to another place. There is even a term for genes that bounce around the genome willy-nilly—they're called "jumping genes." Viruses can also insert sections of DNA into our genomes, and it appears bacteria can too.

Random mutations just tend to happen. When DNA is copied, mistakes can be made. Exposure to sunlight can mutate our DNA. Radiation and a few chemicals can too. Mutations that popped up in the past have led to many of our wonderful modern characteristics, like standing up straight, having big brains, and being great at video games. Some mutations have little value in terms of our species' survival, but have nonetheless stayed in the population (in some cases because they lend reproductive "sexy" value): blue eyes, red hair, not being able to roll your tongue. And sadly, some mutations are quite serious—genetic disorders like sickle-cell anemia, cystic fibrosis, and muscular dystrophy are due to just a few changes in some key genes.

Even a change in one gene and one protein can lead to some startling consequences. A few years ago researchers were baffled when they saw a street performer in Pakistan, a boy of about thirteen, stabbing himself with knives in the hope of collecting tips. It turns out he had a mutation in just a single gene, but this gene was for a protein that worked in nerves. His mutated gene coded for a protein that was so big, it sealed his nerves closed so they couldn't send signals to the brain that said, "Gah! That hurts like hell!"

This mutation has since been found in a few other individuals. Those people have to be insanely careful with their bodies, as they have no way of finding out if they have an injury unless they happen to see it or a bystander points to them and says, "Excuse me, ma'am, your bone is poking through your skin." It sounds funny, but imagine not knowing that you have a horrible gash on your foot, only to notice a day later when it's become infected. Pain most definitely has a vital purpose—reminding us to take care of ourselves, and letting us know immediately when we've failed to do so.

If there ever is an X-Men-like future where mutants and "regulars" are arguing over who deserves rights or who is "natural," we would be wise to remember that truly, we are all mutants. And there's nothing wrong with that. It's why our species has done so well (so far).

We're Not Special

I mentioned in a previous chapter that at the start of the human genome project, most people were sure that we had over 100,000 genes. Surely our amazing brains and wondrous existence could only be explained by having the most genes on the planet. The joke was on us, though. Even grapes have more genes than we do.

Surely then, we have the most proteins of any creature, right? No. Of course not. I'm not saying humans aren't super neato, but we're really not all that physically impressive, and our genome and proteome (the full catalog of proteins we can make) do not get first place in any comparison contest. Don't let this affect your self-esteem, though. I'm sure *you* would win a prize of some sort, such as Most Discerning Book-Reading Tastes.

Moreover, a good part of our genome deals with everyday cellular business that all life shares. Whether you're a bacterium, a mushroom, a pine tree, or an ape, your cells have to get energy, extract wastes, and make proteins. Because of this, a significant portion of our genome is common to most organisms. If you have ever heard the factoid that we share 50 percent of our genes with a banana, it's true. Bananas and we need to do a lot of the same things.

We humans often see ourselves at the top of the world, the pinnacle of evolution, the rulers of the universe. If anything, genetics and comparative genomics can teach us that we really need to tone down our egos. Our DNA has been handed down to us from our ancestors, who if you go back far enough in time, were not human. Our genome does not separate us from other species. Quite the opposite: We are closely tied to the rest of life on this planet by our DNA.

By comparing the DNA of different species, we can see how long ago we shared a common ancestor. DNA is like the ultimate scrapbook of our past—it's all in there. Just take a moment to reflect on the teeny tiny microscopic account of your evolutionary past that resides in each and every one of your cells. Feel the ancient wisdom that infuses your body, and howl at the moon. Or just chill on the couch. Whatever floats your boat.

Junk in the Genome?

We don't have a particularly large genome, we humans. But on top of that, nearly 97 percent of our DNA doesn't even get used to make proteins.

We got this number by looking at the 20,000ish proteins we humans make and tallying up the DNA instructions for them. And the total amount of DNA necessary to make all our proteins only totals to 3 percent of our entire genome. Doesn't seem like much, does it?

When this was first noticed, the other 97 percent of the genome was casually called "junk DNA," since it didn't appear to do anything at all. Once again, you have to appreciate the narcissism that we humans sometimes have: because we didn't know what this noncoding DNA did, someone assumed it had absolutely no purpose and was therefore "junk." Ha. If we don't understand something right away, surely it's worthless, right? Obviously.

Even as a high school biology student, I thought this idea of "junk DNA" was a bit fishy. So as more and more information has surfaced in the last ten to fifteen years about the functions of this noncoding DNA, I always laugh and think, "I knew it!" with just a dash of, "I told you so."

I should mention that not all living things have this surplus of "junk DNA," which I will from here on out refer to as noncoding DNA, since we now know that "junk" is a very misplaced derogatory term. Bacteria, for instance, do not have this noncoding DNA. These superefficient little guys actually use all their DNA. Their genome is much smaller than ours, and they don't have the storage space for huge swaths of noncoding DNA. They don't even have a nucleus in which to house their DNA. It's all exposed out in the main compartment of the cell. I assume this is why the discovery that we have noncoding DNA was met with negativity and assumptions of uselessness. After learning so much about the genetics of bacteria, we probably seemed wildly ineffectual in comparison, having so much DNA we didn't use. Maybe scientists had bacterial genetics envy.

There is another piece to this puzzle. When we say that only 3 percent of the genome has instructions for proteins, we're neglecting stretches of DNA that code for RNA and stop there—that is, the RNA does not leave the nucleus to serve as a guide for building a protein.

It turns out that 85 percent of the genome has genes just for making RNA—about 55,000 different RNA molecules. That's almost three times the proteins we make. Genes that code for these standalone RNA guys are called non–protein coding genes.

What all these RNA molecules are actually doing once they're made is still up for (heated) debate in the scientific community. Some of them might be totally useless. They might get made from a section of DNA and then immediately destroyed. But others could be very important for regulating the rest of the genome.

Indeed, we now know that one of the functions of *noncoding* and *non–protein coding* DNA is the control of the *coding* DNA. When DNA is transcribed to RNA so it is actually used, we say that that gene is being *expressed*. The bits of noncoding DNA between genes actually control the gene expression of the coding, nonjunk DNA. So if anything, this in-between DNA we used to think was useless is in a way the most important part of the genome. The actual genes themselves that code for specific proteins are of course important, but expressing those genes and actually making those proteins at the right time and in the right place is what makes everything work.

Transcription: The process of a strand of RNA being produced according to the sequence of a section of DNA.

Imagine if during development, hormones were expressed in the wrong place or wrong time. I should tell you that some hormones function as a "brain goes here" sign; no joke. So you can see that if that neon sign was in the wrong place or expressed at the wrong time, the consequences could be absolutely disastrous.

Or think about your nails. Those scratching utensils of yours are made of protein, and they should only show up on the ends of your fingers and toes. Just think of how uncomfortable it would be if your hair follicles suddenly decided to stop making soft thin strands of hair and instead started making nails all over your body because the wrong genes were being expressed. Not fun. (This really happened to someone.)

Accounts of medical oddities dealing with improper gene expression could fill a book unto itself, so I won't take you on a *Ripley's Believe It or Not*-ish tour of them, but I think you see why gene expression is just as important as the outcome of the gene itself. Just as in relationships and baking soufflés, timing is everything.

Gene Storage

Believe it or not, your cells actually know how important DNA is (although I use "know" in a very loose sense, as cells aren't so much thinkers as doers) and do what they can to protect the DNA from random mishaps. When your DNA is not being used—and by "used" I mean being copied or being transcribed into RNA—it is stored in a way that offers maximum protection. This protective arrangement for DNA is called a chromosome.

Chromosome: A structure made of tightly wound DNA.

DNA itself is a very thin, threadlike molecule. Step one for getting it into a compact and secure arrangement is wrapping it around a protein called a histone. The DNA winds around it over and over again like thread around a spool. The next step for chromosome building is stacking all of these histones on top of each other in a very pleasing spiral shape. Lots and lots of DNA-wrapped histones like this are what make up the recognizable shape of a chromosome we all know and love.

Chromosomes got their name from the fact that when you apply dye to cells so as to better see them under a microscope, the chromosomes soak up a lot of the dye and are very easy to see. *Chrom* just means color. *Chromosome* in Latin basically means "colored thingamajig." It's not a direct translation, mind you, but it's close. Lots of microscopic things in biology are named by the first schmuck lucky enough to be able to see them, and they are usually named after whatever it most resembles to that one person. Cells got their name because the first guy to see them—Robert Hooke, in the 1600s—thought they looked just like the monks' quarters, called cells, in a monastery. I don't know if he at the time realized he was giving them the name they would always go by. I wonder if he would have given them a more exotic, romantic name if he had been aware of this. Sadly, we'll never know.

Anyway, chromosomes normally are pictured looking like a roughly X-shaped thing, with varying lengths of arms above and below the crisscross point. That is actually what chromosomes look like just before a cell divides, when there is twice as much DNA as usual. The bonus of storing DNA in a chromosome structure is that it is much easier to separate when you're making a new cell, as opposed to having long, tangly strings of DNA flopping all over the place.

DNA is to chromosomes what strands of hair are to dreadlocks. Except, unlike dreadlocks, chromosomes can later unwind to expose the individual threads of DNA. (I've never had dreadlocks, but my understanding is that it's a destination there is no coming back from without scissors.)

Here's the rough schedule of DNA inside of a cell:

» We begin when the cell is going about its regular workday, making proteins, metabolizing sugar perhaps to get an energy boost, and ridding itself of wastes. Fun stuff like that. When this is going on, the DNA is inside the nucleus, looking like disorganized DNA vomit.

» After a while, the cell decides that it's time to make a copy of itself. Almost all of your cells do this; some more often than others. Rapidly dividing cells can be found in tissues like skin and the walls of the intestinal tract. But before a cell can split in two, it needs to make a full copy of the entire DNA manual so that both cells after the division have a copy. It starts to replicate every last strand of DNA.

» When this is done, it starts to condense the DNA into the chromosome shape. Since we right now have twice the amount of DNA as normal, each chromosome has two identical arms that are joined in the middle.

» The cell is now getting ready to divide. The wall of the nucleus (called the nuclear envelope) starts to dissolve away, and the chromosomes line up in the middle of the cell. The identical chromosome arms are then pulled apart and go to opposite sides of the cell.

» The cell then pinches closed in the middle to give you two identical new cells that each get one full set of DNA instructions. At this point they each have DNA in the chromosome shape, but they don't look like the X anymore. They're just chromosome spaghetti.

» For these two new cells to resume their cellular business, the DNA relaxes and unwinds from the chromosome shape so that the DNA can be transcribed into RNA, which will allow the cell to make proteins and be its celly self.

The process I just described is something called mitosis. It's what cells do to make more of themselves. Everything from bacteria to a full human does this at some point or another. Not all cells do it all the time, though. You may have heard that nerve cells aren't super great at making more of themselves, and if you lose 'em, the body doesn't repair 'em. But the vast majority of your cells are doing this all the time. They really enjoy it.

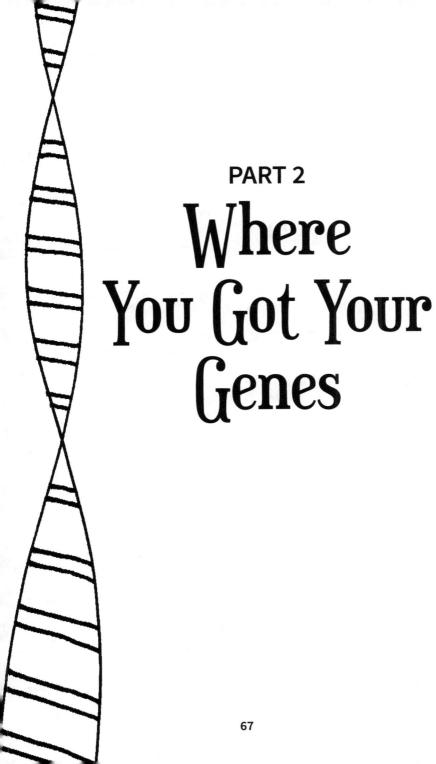

PART 2
Where You Got Your Genes

Let's Talk about Sex . . . Cells

There is another way that your cells divide, but it's not to make more cells for *you*, it's to make a new cell for a *baby*. You probably remember from sex ed, and lots of quasi-inappropriate jokes, that men make sperm and women make eggs. To make sperm or eggs, depending on whether you are a male or female, there is an interesting challenge. You need to give these sex cells, also called gametes, just half of your genome. That way, when the sperm and egg meet, together they have one whole genome. One half plus one half equals a whole (you did it again, math!). But you can't give just any old half. Gametes need to get exactly one copy of every single gene necessary in the human genome. We all actually have two copies of every gene. The gene for the skin pigment melanin—you have two copies of it (unless you're albino, that is). The gene for being awesome (a gene I just made up)—you would have two copies of that too. The genes that determine your eye color—you have two copies of every one of those. It sounds a bit redundant, and it can be, but it's also a great insurance policy that often is helpful. If you get a bum gene for insulin, maybe the second one you get will be functional.

Now, when I say "copy" and "duplicate," I don't actually mean an *exact* copy, which is misleading, I realize, because *copy* ordinarily means exactly that. I apologize for this, but ask that you not blame me for this fun quirk of scientific word choice. By *copies*, I mean issues, versions, or entries—whatever you'd like to call it. When I say you have "two copies" of a gene for awesomeness (which is not a real gene, sadly), I'm saying you have two records in your DNA that relate to awesomeness, but they may or may not be the same.

You have a duplicate set of DNA because you get a whole set from your mother, and an entire set from your father as well. How they gave that to you is in those sex cells. And here's how those are made.

Let's start with a regular cell of yours that has your entire genome that has been completely copied in preparation for dividing, so there is actually twice as much as usual. The DNA is condensed into easy-to-handle chromosomes. At this point, each chromosome looks like that lovely X shape, because each arm has been copied, and they are joined together. The chromosomes line up in the cell, with similar chromosomes next to each other. For instance, your mother's chromosome 15, which has some genes for eye color on it, would be lined up next to your father's chromosome 15.

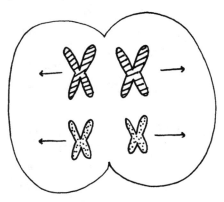

These pairs of chromosomes get split apart and pulled to opposite ends of the cell, which then splits in two. These two cells have just one copy of each chromosome now, but two identical arms of said chromosome. These then line up in the middle of each of these cells again, and this time, the arms of the chromosome get pulled apart. Then the cells once again split themselves in half.

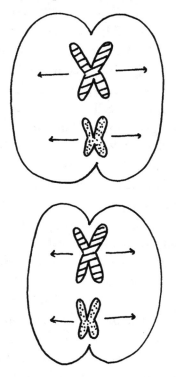

After all of this, we have four cells that each have half the usual amount of DNA. They have just one copy of each gene that you need.

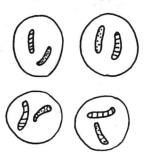

When the sperm and egg meet, the new cell they form, called the zygote, inherits two copies of each and every gene it requires. If all goes well, that new cell starts dividing in half over and over again, and nine months later, it has grown into a complete monster—er, baby.

That whole process of dividing twice to make those sperm and egg cells is called meiosis (pronounced my- OH-sis). It sounds a lot like mitosis because they both involve cells splitting apart, but in meiosis you get four different cells with half the regular amount of DNA, whereas mitosis gives you two identical cells.

Mitosis happens all the time all over the place—on your face, in your stomach, in your butt. Meiosis, on the other hand, only happens in two spots: in the testes for dudes and in the ovaries for the ladies. Meiosis is only for making baby ingredients.

The only kind of DNA that you don't get from both Mom and Dad is the DNA that comes with your mitochondria. Mitochondria are small organelles in your cells that manufacture adenosine triphosphate (ATP for short), which is what cells use to store and spend energy. It's like the cell's gasoline, as it is a chemical that has energy stored within bonds that can be broken and the energy used within the cell, just like the bonds between carbon atoms in long chains power our cars.

The mitochondria you have in your cells came completely from your mom, as only the egg has mitochondria. The sperm doesn't bring any mitochondria to the table, as it would probably weigh them down; in the race to fertilize the egg first, you need to pack light.

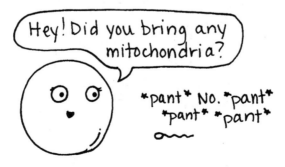

Because the mitochondrial handoff is completely matrilineal, it can be a useful tool for determining ancestral relationships. Our regular DNA is random and mixed together when we pass it to our children, like a hand-me-down set of mismatched dishes. You got half the dishes from Mom and half from Dad, half of which came from their respective parents. And if you have siblings, they got half from your parents too, but a different half. It can make your brain hurt a little bit.

But mitochondrial DNA isn't split up and blended—it just gets handed off in one complete unit, like a single heirloom gravy boat getting passed down from grandmother to mother to you. (Now you can host Thanksgiving dinner! You're such an adult.)

The differences between your mitochondrial DNA and your mother's are very slight, as they would only have come from random mutations you've accumulated during your life. Unless you've been exposed to an unhealthy level of radiation (in which case, go see a doctor), they're pretty much the same. They're really the best present your mom ever gave you.

One Monk's Boredom

The actual distribution of genes via meiosis does have some rhyme and reason to it. In fact, the idea that you got half of your genes from your mom and half from your dad probably doesn't sound all that earth-shattering to you. But there was a time when people had this process all wrong.

Once upon a *long* time ago, thousands of years ago, people didn't associate sex with reproduction, so it appeared as though women spontaneously and independently created babies. There was also a time when people knew men were contributing something to babies, but more along the lines of assistance or some sort of general essence via their sperm.

In some cultures, sperm was thought to aid in the development of the baby, such that if you stopped having sex as soon as you got pregnant, the baby would cease growing.

Elsewhere they thought that having sex with multiple men during pregnancy would give the baby a mix of traits from each of them, as if all their good qualities were blending together. To give their babies the best of all worlds, women would have sex with as many men as possible to give the child all their best characteristics.

When we figured out that semen actually had a distinct purpose, we overcorrected a tad: people thought that women were merely the vessels in which babies grew, but that all the information required to create life was contained within the sperm.

One scientist looking at sperm under a microscope for the first time even described seeing a teeny tiny man inside the sperm, which of course was evidence that the sperm was responsible for the eventual human being. Good job, guys.

Luckily, science was able to scrap the Teeny Tiny Man theory, in part thanks to Gregor Mendel. The father of modern genetics, Mendel lived in what then was Austria (and what is now the Czech Republic) in the mid-1800s. He was an Augustinian monk, and it seems he channeled all his sexual frustrations into gardening. He bred pea plants with the care and consistency only someone with obsessive-compulsive disorder could have maintained. What he showed from this work was that traits were not blended together as was previously suspected, but that they were shuffled and inherited in distinct units, which we now know are genes.

His experiments involved pollinating pea plants himself so he could keep track of the plants' family trees (hooray for puns). He had to make sure that the plants didn't get pollinated the way they normally would, by bees or the wind, so he tied little bags over all the flowers so that the pollen of one plant couldn't get to the flower of another plant without his approval. Yes, the pea plants were his sexual slaves. But in a totally normal way.

He kept track of multiple generations of pea plants and took notes on features like pea pod shape, pea color, flower color, plant height, and even pea wrinkles. He was a very industrious monk with a lot of time on his hands. And a lot of peas.

Here's an example. He had two groups of pea plants: one group all had green peas and one all had yellow peas. He manually cross-pollinated them, which means that he took a little brush, picked up pollen from the green pea plants, and dabbed it onto the flowers of the yellow pea plants (or the other way around). In this way, he knew who the plant parents were for all the resulting baby plants (which is in no way a real term, but *offspring* sounds so detached and boring to me). When the flowers eventually produced seeds, he planted those and waited for them to grow.

When they grew into the wonderful pea plants he so loved, he noticed that every single plant had green peas, despite the fact that the parent plants were half green and half yellow pea plants.

He then kept track of this new generation and cross-pollinated them with each other, using the same flower brushing and dabbing technique (which is surely what he called it too).

Artificial pea plant insemination.

I point out the actual method for cross-pollination that he used because when I took biology in high school, I remember always hearing about this "cross-pollination" he was doing, but I didn't fully understand what that meant. The visual of this monk gingerly dabbing pollen from one flower onto the pistil (the female part of the flower) of another makes it real for me, as I hope it does for you.

When he raised the resulting seeds from this cross, 75 percent of the plants had green peas, but 25 percent had yellow peas. Traits that can disappear for a generation only to reappear this way, he dubbed recessive traits. He realized that they were not disappearing entirely, but were being covered up by what he called a dominant trait. Well, I'm sure the Austrian term is different, but you get the point.

Dominant: What we call a form of a gene when it gets expressed and can cover up another form of the gene (since we all have two copies of each of our genes).

Recessive: An allele that can get covered up by a dominant one. A recessive trait is only expressed if you receive two copies of it.

Are you getting a sense of how much painstaking work this was? This guy spent years and years keeping detailed logs of plants and their entire ancestries, keeping notes of all their traits. This is stuff that only monks have time for. I'm just glad he didn't choose a different hobby like basketweaving or forced perspective chalk art.

After noticing that some traits were dominant over others, he worked out the explanation and described the rules for genetics that we still talk about today. We call it Mendelian genetics because we love him so much.

Mendel's two big ideas are now called the law of segregation and the law of independent assortment.

The law of segregation sounds terrible. Let's just get that out of the way right now. But I promise there is no racist tint to this law of genetics. In fact, what this law actually describes is probably going to sound pretty obvious to you, since you have been living in the twenty-first century. But at the time, Mendel explained that when pea plants (or guinea pigs or humans) reproduce, they make sex cells (eggs and sperm) that have half the usual amount of the genetic information in them. You don't just get any random half, but rather information about each specific trait, which meant that the factors that caused those traits had to be separated out accordingly when those sex cells form. The separation part is what led him to refer to it as segregation.

An important part of this law is that it takes into account that you have two copies of each of your genes, and when you make sperm or egg cells, you put just one copy of each gene in there.

The other law, independent assortment, Mendel explained, meant that genes aren't linked to one another. Just because a parent plant was tall with green peas doesn't mean that being tall and having green peas will necessarily be inherited together. Because genes are segregated randomly when sperm and eggs are formed, traits are inherited independently of one another.

We know now that this rule actually is not *always* true, which is sort of a theme in biology. Yes, traits are inherited as distinct units, but there are indeed cases where having one trait makes you more likely to have another, such as being male and bald. And why would that be? Because the genes for these traits are on the same chromosome, so they can't be segregated according to Mendel's first law. He lived in a time way before we knew anything about DNA, genes, or chromosomes, so we will not hold this against him.

Like many people who were way ahead of the pack, Mendel died long before anybody gave a hoot about all his groundbreaking work. But he sure is famous now, and I'm really glad that he liked gardening so much.

When Genes Duke It Out

All genes are not created equal. They come in different flavors, and some are better than others. Well, not necessarily *better*, just more functional (which is usually better). Because as it turns out, the genes you get, and what actual traits you wind up with from those genes, are slightly different.

First of all, those gene flavors I referred to have an actual name: alleles. Alleles are the different varieties a gene comes in. Let's consider Mendel's precious pea plants. There is a gene that determines the color of the flower on the plant. The different alleles for flower colors are white and purple.

Allele: A form of a gene. The gene for flower color, for instance, may have a pink allele and a purple allele.

Now if *alleles* is one of those intimidating-sounding terms to you, like *quantum physics*, just mentally substitute *flavors* in there each time you see it. Actually, I'll do it for you. Instead of saying alleles, I'll say flavors, because that sounds like fun.

Alleles of ice cream.

But getting back to the point at hand—the different flavors of those genes are what account for the amazing variety among people, not to mention the rest of the tree of life. (It's not all about us, after all.)

But you can't necessarily tell what flavors of genes a person or any other living thing has just by looking at them. Some genes can be hidden from view because gene flavors can be dominant or recessive. And in between that there is codominance and incomplete dominance. I know, I said I wouldn't term you to death, but these are some good ones.

Dominant vs. Recessive

Let's pick these apart, starting with the dominant and recessive hoopla. If a certain gene flavor is dominant, it doesn't mean it's bossy and controlling, or that it's into whips and chains.

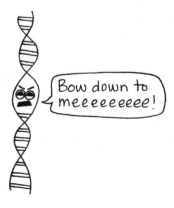

Bow down to meeeeeeeee!

It just means that if you have this gene flavor, you are definitely going to express it. The alternative is being recessive, which means that the gene flavor can be masked. In the example of pea plant flowers, purple is dominant. If you get the gene for purple flowers, you express it.

Remember that you (and pea plants) are issued *two* copies of every gene, which may or may not be the same. But even if you get one gene for purple flowers and one gene for white flowers, that pea plant is going to make purple flowers because the purple flower part is dominant. What makes it dominant? Ordinarily a gene flavor is dominant because the trait it produces is due to a protein it codes for, and the recessive trait is what you see if the protein is either not made or does not work quite right. This could happen if the recessive gene is a mutated or otherwise nonfunctional gene.

For the pea flowers, the gene that causes purple flowers has the code for a protein that gives those flowers that lovely color. The recessive gene doesn't code for a protein that has pigment, so if you get two of these gene flavors, you don't get any color in the flowers, and they're just plain white.

The thing about showing a dominant quality like purple flowers is that you don't know if that plant has two genes for purple flowers, or if it has one gene for purple flowers and one for white flowers. The only way to tell is to do what Mendel did, which is breed that purple pea plant with a plant with white flowers and see if any of the pea plant children get white flowers. If there are pea children that have white flowers, it means that the purple-flowered parent was smuggling a recessive white flower gene under its jacket. When this gene met up with the recessive gene from the white-flowered parent, you got a white-flowered baby pea plant.

So sometimes genes are dominant, and if they're present, they manifest in the actual traits. If genes are recessive, you only see the effects if you have *two* of those recessive gene flavors. Look at all this learning we're doing together.

While this whole "dominant" and "recessive" gene stuff does explain how a trait can skip a generation—because a trait can be recessive and be masked—it doesn't actually explain the way that people commonly talk about traits that "skip a generation." There are a lot of myths out there about traits that hop, skip, and jump generations, and the truth is usually much more complex, such as the likelihood of having twins, being gifted in the arts, or being a crazy weirdo. These sorts of things aren't due to the kind of single gene cover-ups that can happen because they're recessive. Sorry to disappoint.

Codominance and Incomplete Dominance

Codominance happens when you actually express all of the gene flavors you get. There is no covering up or hiding. A good example of this is blood types. If you get both A and B genes from your parents, you are blood type AB. You get both of them. They don't cover each other up. They're both equally bloody.

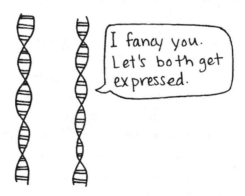

Blood type O is still recessive, though. There actually isn't a gene for this O; you are said to have blood type O if you don't have genes for A or B. It's a blood type of exclusion, really.

Incomplete dominance is when there are three possible outcomes from the gene flavors you get. Think about snapdragons. Don't argue with me; just do it! In snapdragons, there are two gene flavors for the color of the flower: red and white. If you have two red genes, you are red; if you have two white genes, you are white. But, if you have one red gene and one white gene, you are (drumroll, please) *pink*. In this case, you don't get the usual dominant-recessive situation where one completely takes over. If you have one of each, the actual physical trait is somewhere between the two extremes. It's a mighty good thing that our buddy Mendel didn't work with snapdragons. If he had, he might have been ultra-confused about all this stuff.

The Magic 46

All your genes, whether dominant or recessive, functional or mutation-laden, are hanging out in your chromosomes. You (probably) have 46 of those chromosomes. Most people do. You also probably got 23 of those from your mother and 23 from your father. (I'll talk about exceptions to this statement a little later.)

Chromosomes don't have names. I can't decide if this is a good thing or not. If all of them had names, it would be an awful lot to memorize, but it might be more interesting than what they have now—numbers or letters. There's chromosome 1, 2, 3, all the way to chromosome 22, and then you have the X chromosome and the Y chromosome—the sex chromosomes.

So what is going on with all these chromosomes? What information do they carry? How big are they? What do they look like? What TV shows do they watch? Let's start out with the numbered chromosomes, 1 through 22. We'll talk about X and Y later, when I've had more coffee.

Of the numbered chromosomes (which are also called autosomal chromosomes), numero uno is the biggest, with about 3,000 genes, and 21 is the smallest, with a mere 400 or so genes.

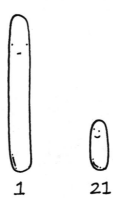

1   21

Chromosome 21

Let's first jump to that little chromosome 21, because it's among the most talked-about chromosomes. If a person ends up with three of them, it's called Down syndrome.

How would someone get three of these chromosomes, when you're supposed to just have two? This can happen when during meiosis—when cells split apart to make sperm and egg cells—the chromosomes get stuck together and don't separate the way they're supposed to. When this happens, you can wind up with an egg that has two number 21 chromosomes. Later when it meets up with a sperm that also contributes a chromosome 21, you end up with a total of three.

Having extra chromosomes like this happens all the time, but if it happens with any numbered chromosomes (as in, not X and Y) besides number 21, it usually results in a miscarriage, as development goes completely haywire because of the extra chromosome. It seems that as far as having three chromosomes goes, an extra 21st chromosome is the only case that results in a viable baby, probably because there are so few genes on that particular chromosome.

Which Genes Are on Which Chromosome

So what are the rest of the chromosomes up to? Researchers have been hard at work making chromosome maps for you, pinpointing where each gene lies on these chromosomes and what it does.

That's it! Right there!

A lot of work has been put into finding the homes of genes that cause diseases like Alzheimer's, diabetes, and cancer. In fact, now that we know the addresses of lots of scary genes, you can get tested to find out if you have them, and what your odds are of at some point developing, say, breast cancer.

Now, I'll go over some highlights of the human chromosome's contents. But keep in mind that complicated conditions like epilepsy, bad eyesight, and many types of cancer have related genes on multiple chromosomes. I'm also listing quite a few diseases and oddities, because those are the sorts of genes that researchers located first. After all, a great deal of medical research concerns figuring out what causes things to go wrong. But please don't interpret this list (which contains many mentions of diseases and disorders) to mean that your genes are just rife with horrible DNA mutations that could kill you. I'm just saying that this is where certain variations of genes *could* be—not that you necessarily have them. I'm sure you're fine. Your DNA loves you.

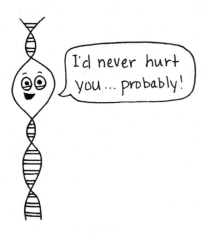

» Chromosome 1 is the home of genes that are associated with deafness, schizophrenia, and something called (I swear I'm not making this up) maple syrup urine disease.

» Chromosome 2 has genes for red hair, nearsightedness (hey, I have that!), and muscular dystrophy.

» Chromosome 3 has some of the genes for breast cancer, night blindness, and being very short.

» Chromosome 4 has genes for alcoholism, susceptibility to psoriasis, and Parkinson's disease.

» Chromosome 5 contains genes for ADHD, taste receptors, and dwarfism.

» Chromosome 6 has genes for celiac disease (as in, people who go on gluten-free diets out of medical need, not just because it's so cool these days), epilepsy, and dyslexia.

» Chromosome 7 is a fun one: it has genes for obesity, missing fingers and toes, as well as extra fingers and toes.

» Chromosome 8 has genes for scurvy, liver cancer, and congenital adrenal hyperplasia (which is explained in the next chapter).

» Chromosome 9 has genes for melanoma, albinism, and susceptibility to lead poisoning.

» Chromosome 10 has genes for type 1 diabetes, prostate cancer, and severe combined immunodeficiency disease, sometimes called "bubble boy disease."
» Chromosome 11 contains genes for dopamine receptors (a very pleasurable neurochemical), lung cancer, high bone mass, and in a strange twist, also osteoporosis (which is the opposite of high bone mass).
» Chromosome 12 has genes for alcohol intolerance, Alzheimer's disease, and a propensity toward bedwetting (nocturnal enuresis).
» Chromosome 13 has genes for cataracts and pancreatic cancer.
» Chromosome 14 has genes for defenders against cell death and DNA mismatch repair—that one could definitely come in handy.
» Chromosome 15 has genes for brown hair and brown eyes.
» Chromosome 16 has genes for vulnerability to UV-caused skin damage, kidney disease, and allergy susceptibility.
» Chromosome 17 has genes for ovarian cancer, serotonin transporter (which is related to anxiety), and dementia.
» Chromosome 18 has genes for carpal tunnel syndrome (which I think of as an office allergy), Paget bone disease (abnormal bone growth), and colorectal cancer.
» Chromosome 19 has genes for green and blue eyes, cleft palates, and susceptibility to malaria.
» Chromosome 20 has genes for fatal familial insomnia, gigantism, and anemia.

Fatal familial insomnia is a very rare genetic condition that causes a very intense form of insomnia. It starts in a person's thirties or forties usually, and it is caused by a very strange mutant protein called a prion. Prions are nasty fellows that go around and make other proteins inactive. In this disease, it wreaks havoc on the brain, and eventually the brain becomes physically incapable of entering a sleep state. It is, as the name of the disease suggests, eventually fatal, but the whole process takes years. There are only a handful of families in the whole world who have it, so this is one thing you can cross off your list of worries. If it were in your family history, you would definitely know.

» Chromosome 21 has genes for influenza resistance and, as previously discussed, Down syndrome.

» And last but not least, chromosome 22 has genes for a wide range of rare diseases that I have quite frankly never heard of, but most of them are fatal and/or affect brain development. Not fun stuff.

Once again, I'm not saying that all of your chromosomes, and all your friend's chromosomes, have these particular genes for these particular traits. This is just where, in individuals who have these characteristics, the genetic cause was found. And keep in mind that we don't have the complete map of all of our chromosomes. We're close, but we don't have all the addresses of the residents just yet.

It can also be confusing when we call genes by the disorders they can cause. The breast cancer gene, for instance, is a gene that normally *prevents* breast cancer. People that have a certain form of it (with a mutation) are predisposed to breast cancer because the gene is *not working*. So discussing whether we have gene BBBB (which stands for Blah Bleh Blah Bleh) isn't meaningful. Since we're all human beings, we all have the same genes (with rare genetically odd exceptions). The question is if the genes are working or not, or if they have mutations that make us vulnerable to cancer, genetic diseases, and other such fun results.

When we know everything about what's happening on each of our chromosomes, it will spell happy times for genetic testing companies. Imagine if you could easily find out which of these gene varieties you have. Some people would likely find this information helpful and interesting, while others might find that it causes fear and anxiety. Such is the paradox of scientific progress, I suppose.

The Sex(y) Chromosomes

We still have to talk about those lettered chromosomes, the X and the Y. What is happening on these particular chromosomes? For starters, the X chromosome is far bigger than the little nub that is the Y chromosome. Madame X chromosome contains about 1,400 genes; Mr. Y has a mere 200.

The products of the Y chromosome genes are almost exclusively manly hormones like testosterone that produce masculine traits like a hairy chest, a hairy face, and a hairy ass crack. Just a lot of hair, really.

The X chromosome, in contrast, is just teeming with vital information, like the genes for baldness, colorblindness, and about 1,398 others.

So what do those sexy sex chromosomes do? They have that name, of course, because they determine that rather hefty part of our identity—whether or not we are male or female. The prevailing wisdom is that XX is a girl (like myself), and XY is a boy. Right? Actually, not always.

With biology, there are very few things that are true all the time. Life is messy.

For starters, not everyone gets two sex chromosomes. Some people wind up with extra ones, or are missing one. You can have XXY, XYY, XXX, or just X. Some of these states are associated with a high likelihood of mental retardation, but some, like XXX, could be completely unknown—that is, until the person tries to have a kid. If you have the wrong number of sex chromosomes, you're probably sterile.

But let's say someone does have two sex chromosomes—either XX or XY. YY, we can pretty safely assume won't be happening. Y would only be in a sperm, and two sperms don't, as far as I'm aware, combine to make babies. That would be interesting, though.

But even if you are XX or XY, sex doesn't always manifest based on which two chromosomes you have. It's about hormones, a complicated balance of hormones. And this can get tricky.

A person can be XY and appear female or be XX and appear male.

A person with XY chromosomes can appear female with a condition called androgen insensitivity syndrome. The Y chromosome has the gene for testosterone, which gets produced just as it does in many other guys, but the *receptors* for testosterone don't work. So even though testosterone is present, it can't do what it normally does: make a penis, put hair on the chest, and help with that manly physique. So someone who has this condition grows up appearing female, and often only finds out at puberty when no menstruation happens that she actually doesn't have a uterus or ovaries.

A person can be XX but appear to be a male because of congenital adrenal hyperplasia (which I mentioned in the previous chapter), where hormones in overdrive during development masculinize the body even though there is no Y chromosome. In an extreme case, a baby boy could be born and mature normally, having a fully functioning penis (and sex life), only to find later that not only does he have a penis and testes but, internally, a uterus and ovaries as well.

In fact, the realization that this is the case is usually preceded by the individual experiencing abdominal pain due to menstruation caused by this hidden uterus. With no external exit strategy for said menses, this can be quite painful and dangerous.

And finally, there is an issue of gene crossover. A female, XX, can still have a gene from the Y chromosome, such as the one to produce testosterone, because during meiosis in sperm formation, the X and Y chromosomes are near each other, and genes can be swapped across chromosomes. Does this make the woman who is XX and has female body parts less of a woman just because she has a gene to produce testosterone? The Olympics thinks so.

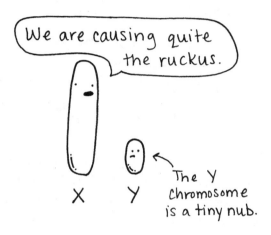

Sex, it would seem, is not the binary system we think it is. And our way of determining male vs. female is not the only way to do this:

» In some insects like grasshoppers, crickets, and roaches, there is only one kind of sex chromosome, X. If you get two of them, you're female, and if you just get one, you're a male.

» In birds (and even some insects and fishes) it's the females that have different sex chromosomes. Females are ZW and males are ZZ. Scientists changed the letters to differentiate between the

XX-female, XY-male scenario in humans. That was so thoughtful of them.

» In ants and bees, there are no sex chromosomes at all. Instead, sex is determined by whether or not an egg was fertilized. If the egg isn't fertilized, the offspring is male. If the egg is fertilized, it's female. So male ants have no fathers, and they have half as many chromosomes as females. The poor little dears.

Depressed ant

is depressed.

My favorite system for sex determination is in a species of deer called the Indian muntjac (as opposed to the Chinese muntjac, of course). They have *three* different kinds of sex chromosomes: X and two versions of Y, called Y1 and Y2. Females are XX, and males are XY1Y2. Ha, those crazy muntjacs!

May we never forget that biology is a complicated, messy, random process that has found several ways for dealing with sex and reproduction (and just about every other bit of life business). There is really no rhyme or reason for any of it. If it works, it works. If it doesn't, you die. That's pretty much the only rule in nature that applies all the time.

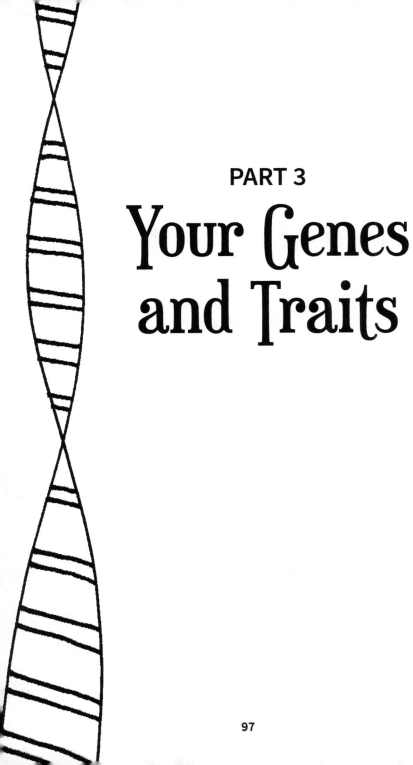

PART 3

Your Genes
and Traits

Sometimes One Is All It Takes

There aren't many noticeable traits that are controlled by just one gene. Usually anything that's worth a damn is also relatively complicated. Even the fur color of a Labrador is controlled by two genes, for Pete's sake.

If a trait is controlled by one gene and has a few (usually two) possible outcomes, we say it's Mendelian, because it is just like Mendel's precious pea plants' height, flower color, and pea wrinkliness.

Classic examples of Mendelian inheritance in humans are things like:

» tongue-rolling (that is, being able to or not being able to)
» attached or free earlobes
» having a widow's peak hairline or not
» having hairy or not hairy knuckles

I was planning to cover all of these classic examples until I started researching them further and found that they are all, in fact, quite false. It turns out that each of those examples is actually controlled by multiple genes, not just one.

I cannot tell you the betrayal I felt when I discovered that every example I've learned about and held dear since freshman biology with Mr. Worthen at Reno High School has been a bald-faced lie. When the reality of the falsehood truly set in, it was like the end of *Sixth Sense* when Bruce Willis realizes he's dead. Utterly traumatizing.

Even eye color was once thought to be Mendelian—that the two possible outcomes were blue or brown. That myth bit the dust quite a while ago. After all, you don't have to stare longingly into too many people's eyes to see that there is quite a bit more variation between eye colors than just blue and brown. I, for instance, have greenish eyes, which was previously explained away as "just a shade of blue." Just a shade of blue? How dare you. I'm *special*, dammit!

And so it's true of these other so-called classic examples as well. And it makes sense. How could just one gene coding for one kind of protein make the difference between attached earlobes and free? A straight hairline and a widow's peak? Hairy knuckles and baby's bottom knuckles? It doesn't make any sense.

Nope.

These erroneous examples were even in my old (2002) biology textbook, the otherwise trusty Campbell *Biology* book that is common to most college courses. This is the trouble—if you want to call it that—with biology and genetics in particular: we're discovering new things every day, so textbooks fall out of date very quickly. These examples of Mendelian traits in humans have particular staying power, though. I'm sure that at this very moment, a well-meaning biology teacher is asking a whole classroom full of students to roll their tongues, to soon explain that those who can't do it have two recessive alleles for a mythical tongue-rolling gene.

So now that you are one of the most knowledgeable people about the false examples, you can correct anyone who talks about the genetics of earlobes, tongues, and widow's peaks. Until a lot of textbooks are phased out, this myth will surely continue. The only common example that seems to be accurate is a gene that determines if your earwax is wet or dry. Yeah. That's really all I have for you. Earwax.

What is this amazing earwax gene? It comes in two flavors: wet and dry. If you have wet earwax, it's on the yellowy side and a bit sticky, and if you have dry earwax it's crumbly and tan. I am a wet earwax person, so I can't quite imagine having the dry version. You may be the opposite. The gene flavor for dry earwax is recessive. The actual reason for the difference appears to be that the recessive gene flavor codes for a protein that is not as effective at making the earwax, well, waxy.

Another fun side effect of your earwax is its impact on body odor. If you have wet earwax you are also prone to stinky armpits, because another effect of this gene is that you secrete proteins from your sweat glands that bacteria find extremely appetizing. They eat the proteins and thrive in the moist comfort of your pits, and their waste products stink it up.

If you have dry earwax, you don't have this BO problem (and I am very jealous of you). There isn't much bacteria food on your skin, so you don't get a bacterial party and the associated odor. Congratulations to you.

The gene for dry earwax and barely-there body odor is recessive, but it's very common in parts of Asia. I'm guessing that they don't have as many commercials for intense deodorants and their "heat activated," "pH balanced," and "clinical strength" qualifications in these areas.

So when it comes to random variety and everyday traits like your face, your eyes, and your hairy knuckles, one gene does not control what you get. Yet there are traits that are definitely inherited in a Mendelian manner—with dominant and recessive alleles (flavors) of just one gene. There are lots of examples of that, but most of them result in debilitating diseases, genetic disorders, and cancer (which we'll get to later)—all because one of the varieties of those particular genes codes for a protein that doesn't quite do the job it's supposed to do. Compared to all these, earwax sounds like a dream.

A Gene Symphony

Your best qualities (besides your first-rate earwax, of course) are due to a symphony of genes working together to bring about a certain trait. Your height, the shape of your nose, and your exemplary yo-yo skills cannot be explained by Mendelian genetics.

For things to be Mendelian (due to one gene with two gene flavor options), the traits we see in all of us humans would have to take just two forms. But clearly things like height, talents, personality, and comedic prowess exist as a broad continuum, not distinct all-or-nothing states.

But I like thinking about hypothetical situations, so let's consider a world where all of our characteristics were inherited in a Mendelian fashion.

If we were like pea plants, people would either be six feet tall or three feet tall. They'd either have white hair or black hair, pale blue eyes or dark brown eyes, and nothing ever in between. People would be either comedic geniuses or humorless zombies. They would either have Einsteinishly high IQs or be completely intellectually hopeless. The world and our lives would be simultaneously extreme yet predictable and rather boring.

But shopping for pants would be pretty easy, I suppose, as there would only be two sizes. And in school, there wouldn't be a bell curve, but rather a two-peaked score distribution with nothing in the middle.

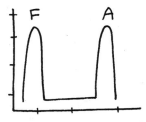

The double bell curve.

Luckily (and obviously) our most interesting features are not all-or-nothing, but have a wide range and exist in many combinations. And very few traits are related to another. Being tall doesn't mean you'll be a good writer. Having a big head doesn't mean you'll be smart—although I'd like to think my giant skull means I have a massive brain. Yes, I really do have a giant head, at least as far as people who size girls' hats are concerned. They never fit on my noggin. My life is so hard.

Anyway, these interesting characteristics exist in a seemingly infinite number of possibilities because there are so many genes that contribute to them. Let's think about the math for just a moment. Let's pretend there was one gene that determined whether a person is smart or stupid, and let's say that this gene exhibited incomplete dominance (oh, if only being smart were dominant!). This means that someone who got two smart genes (one from each parent) would be smart, someone who got one of each would be average, and people who got two stupid genes would be—well, you know. So two genes might give us three possible outcomes.

Now let's pretend there are *two* genes that determine whether you're smart or stupid. Now you could get four smart genes; three smart, one stupid; two smart, two stupid; one smart, three stupid; or four stupid. That's five different possibilities. And it keeps on going up from there. So if you think of characteristics like general intelligence, height, or shape that have a broad range of outcomes in very small increments, you know that there are a lot of genes at work to make these traits happen.

And let's not forget that what your genes dictate is a theoretical maximum or potential ceiling, but the actual outcome depends on what happens to you in life. Someone who has genes for tall stature, for example, might not reach his full potential height if he is wildly malnourished as a growing boy. But even then, he still possesses those genes for tallness, so he can pass them on to his children.

We'll dive deeper into the interplay between your genes and your environment later, because there is a whole lot going on there. For now, just remember that your more interesting traits are due to a whole slew of genes that work together to bring them about.

Unfortunately for us, working out the functions and relationships between dozens or hundreds (even thousands) of genes is as complicated as the traits themselves. It will take a very long time to have the full encyclopedia of genetics done. And right now, we're working hardest on gene families that contribute to serious diseases like cancer because that is where we have the most incentive to unravel the mysteries. The genetics of less serious (but certainly no less interesting) details like personality, hair color, and toe shape will probably be among the last to be solved, since our priorities lie in helping people and curing diseases (as they should).

But really, let's hurry up and cure all those genetic diseases so we can figure out what genes make my head so big. I don't have all day.

Taste Receptors

You probably don't think about genetics when you take the first chip from a pile of nachos or ponder the spinach salad at the local gastropub (what, you don't ever do that?), but there is a part of your genetic self at work here—the genes that contribute to your taste receptors.

The reason some people like to eat broccoli and others want instead to clock you over the head with a floret of it, and the reason some people can't get enough spicy food while others feel like hot sauce singes their mouth with a flamethrower? People have different taste-related genes.

It turns out that not everyone tastes alike—I mean, everyone's *ability* to taste is not alike. I'm actually pretty sure that we all taste the same if eaten. Not that I would know from experience. I'm many things, but definitely not a cannibal.

A quick example is cilantro. For some, like me, cilantro tastes fresh and grassy, making it a perfect accompaniment for salsas, salads, and maybe some gazpacho. For others, cilantro tastes like soap, which is ordinarily not a desirable taste. Whether or not you enjoy the taste of cilantro depends on a single gene.

Allow me to channel my favorite infomercials and say, "But that's not all!" as far as the exciting differences in people's tasting abilities.

There is a certain group of people that are described as *super*tasters. This fabulous "super" designation was coined by

Professor Linda Bartoshuk at Yale University, who during a study for a sugar substitute, noticed that about a quarter of her taste testers reported a bitter aftertaste.

So what's the difference in these supertasters? Two things: they have more of those taste-receptor-laden bumps on their tongues and have a gene that allows them to taste certain compounds as bitter. People that have a different form of this gene taste nothing at all from these particular chemicals.

It turns out that about 25 percent of the population is made up of supertasters. About 50 percent are medium tasters, and the last 25 percent consists of so-called nontasters. Nontasters do not, as it might sound, suffer from a lonely world of complete tastelessness (although that situation *can* happen if a stroke affects a certain part of your brain), but just are at the bottom of the spectrum in terms of their ability to discern mild tastes. And the sexes are not evenly split among these categories: women are more likely to be supertasters. Make of that what you will.

While I love the term "supertaster," as it conjures up images of tongues with capes rescuing babies from burning buildings (for me, at least), it is somewhat misleading.

SUPER TONGUE

There is nothing terribly glamorous about being able to taste more things acutely and finding bitterness where others don't. If anything—and here's where the *super* designation is most apt—it is a burden to bear. People with less tasting powers, the medium and nontasters, can be filed under the "ignorance is bliss" category.

How can you find out if you are a supertaster? You can order some PTC strips (just Google it; they're even on Amazon) and see if they taste bitter to you. You can also count your taste buds. Not all of them on your entire tongue. That would be tedious as hell, if not impossible. No, here's what you do:

» Get some blue food coloring and head to the bathroom. Make sure no one is around and lock the door, lest someone walk in on you painting your tongue, thereby rendering the whole otherwise normal science experiment a supremely awkward, embarrassing encounter.

» Find one of those reinforcement rings for hole-punched paper. This will provide you with the specific area to count within.

» Take a cotton swab and wipe the blue food coloring on your tongue. This will dye the background of your tongue and leave your little buds of taste (fungiform papillae, to be exact) pink. Put the little sticker on your tongue, and start counting inside the circle. If there are less than 15 pink bumps, you're a nontaster; if there are between 15 and 30, you're a medium taster; and if there are over 30, you're a super-duper taster!

Go find out what you are, taste-wise. I'll wait. Try not to eat the paper hole reinforcer sticker thing.

Mitochondrial Genes

Inside every cell in your body, there are very strange organelles called mitochondria. Although they are helpful and prolific, they're not quite "yours." Mitochondria are actually itty-bitty cellular stowaways. They have their own DNA, set their own schedule, and reproduce themselves whenever they damn well please. They have this level of autonomy because they used to live on their own and pay their own bills. Their residence in your cells is explained by the endosymbiotic theory.

Here's how (my incredibly insane, simplified version of) the story goes:

Once upon a very long time ago, there was a mitochondrion named Manfred. He was having a really great time making the energy-storage molecule ATP, adenosine triphosphate. He adored making this ATP. It gave him a sense of fulfillment and gave his life purpose.

One day, he was floating around as usual when a cell a thousand times bigger than him approached. Manfred cowered in the shadow of this mighty beast. He tried to get out of the way, but it was no use: he was quickly swallowed completely by the monstrous cell.

Once inside, he looked around in terror. No doubt any second an acid-filled lysosome would be by to digest Manfred alive. What a way to go. But as there didn't seem to be one in the vicinity, he began to make ATP to soothe himself. He spewed out a generous dose of ATP, and to his amazement, nearby ribosomes and enzymes happily gobbled it up. They were getting energy from it to do their cellular business. Feeling proud, Manfred made more ATP by using some glucose sugar he saw float by. No one seemed to notice he was there, and the ATP appeared to be useful for everyone around him.

After a few days inside the cell, Manfred began to feel safe—far safer than he had felt in his old life outside, in fact. He then decided to stay in that cell forever and call it home. He made more of himself by splitting in half over and over again. The cell still didn't seem to mind. How marvelous.

Soon though, the giant cell started acting strange. Everything was moving about, and before long the middle of the cell started to pinch itself closed. Half of Manfred's daughter mitochondria were on the other side, and as the cell split in two, he solemnly waved goodbye to them.

Manfred consoled himself by making more baby mitochondria, knowing his offspring in the other cell would do the same. This cell would surely pull that nonsense pinching business again, and he'd be ready for it this time.

The end.

And today, millions of years after Manfred's ordeal, we all have the descendants of these mitochondria in each of our cells, and they still do us a favor by making ATP that our cells can use to make proteins, move around, and excrete wastes (among other things). They're very helpful cellular tenants, those mitochondria.

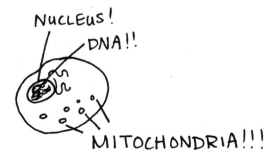

The ATP that they make is a molecule that is made of adenine—the same adenine that's in DNA—surrounded by three phosphate groups—the same thing that's in the side chain of DNA. What the mitochondria are actually doing that's so helpful is attaching a third phosphate to adenosine diphosphates (ADPs) that are floating around the cell. They slap a phosphate on to yield an adenosine *tri*phosphate, which then meanders around the cell and gets used for a wide variety of cellular activities.

By "used" I mean that the ATP is converted back into ADP, and in doing so releases a little burst of energy that can be utilized by the cell. All the sorts of activities I've mentioned so far, things like DNA replication, making proteins, and being awesome, are actually made possible by ATP and the energy it carries.

In this sense, cells are using the greenest of energy technologies, as the ADP is recycled and turned back into ATP by the mitochondria, which then goes back out and helps the cell do its work. It's a highly efficient system with very little waste. Kudos to you there, cells! Now how the hell do we do something equally fantastic to power our cars and heat our homes? Anyone? Bueller?

You might remember from a few chapters ago that the mitochondria you have in your cells at this very moment are entirely from your mother. You didn't get a single one of them from your father. It's not that mitochondria don't think your dad is a standup guy and snazzy dresser; it's just that your dad's sperm didn't bring any mitochondria with them when they raced to fertilize your mom's egg.

Now, like I said before, mitochondria have their own set of DNA. It is quite logically called mitochondrial DNA. (Isn't it great when names make total sense?) You and your mumsy have the exact same mitochondrial DNA, minus the random mutations you and your momma have accumulated since you were born (or, more specifically, since the egg that wound up becoming you got its mitochondria from your mom).

Mutations in mitochondrial DNA happen the same way they happen in regular, nuclear (in your nucleus) DNA: by random mistakes, good old-fashioned chance, and outside influences like radiation, for instance.

Mutations in mitochondrial DNA are often harmless and can be used to construct family trees, since the more similar two people's mitochondrial DNA is, the more recently they shared an ancestor. But those mitochondrial DNA (abbreviated as mDNA) mutations can also cause genetic disorders.

The most common disorder from mutated mDNA is MELAS, which stands for (hold on) mitochondrial encephalomyopathy, lactic acidosis, and stroke-like episodes. Yeah. It's more of a sentence than a term. This condition has such a ridiculously long name because it's inclusive of the many different ways it can manifest. Affected individuals may be shorter than average; they might experience "stroke-like episodes," which can include headaches, vomiting, and visual disturbances; and MELAS can also bring about type 2 diabetes and hearing loss.

Believe it or not, this broad condition is caused by just a single difference in mitochondrial DNA: base number 3,243 is G instead of A. That's all it takes, my friend. Just one little difference throws the entire machine off balance. Mitochondria are just that important: the ATP they make is vital to *everything* you do.

And for the record, I'm not usually comfortable saying "everything" when it comes to biology. There are almost always exceptions, even when just talking about the human body. I can't often say that "all cells" do this or have that, because there is usually an exception. As my high school biology teacher used to say (all the time), "*Always* and *never* are never true in biology." Sure, it's somewhat self-contradicting, but it's also a very good rule.

But in the case of mitochondria, I really do mean that they affect everything you do, because without them your cells wouldn't have enough energy to do much of anything at all. Those little guys are so important. Thanks, mitochondria!

MIGHTY

mitochondrion

Guys, Baldness Has Nothing to Do with Your Dad

In a stereotypical commercial for a hair-growth product that helps men recultivate their bald patches, you might have beheld a scene such as this: a man walks down a hallway practically wallpapered with family photos as he describes the annoyances of baldness. He gets to a picture of his father and sassily says, "Thanks, Dad."

While it's understandable that a man might assume he inherited the baldness gene from his father, it's simply not true. Guys, if you go bald, the only person you can blame is your mother. But don't be mad at her. It's not like she did it on purpose.

How is this Mom's fault? The gene that makes the difference between being bald or having a full Fabio-esque head of hair is on the X chromosome. Because this gene is on a sex chromosome (which are the X and Y chromosomes), it is said to be sex-linked. This means that your odds of getting the resulting condition from a gene that sits on the X chromosome are affected by whether you're a male or female. Those horribly sexist genes.

Now, how does a person get these so-called sex chromosomes? You get one from your mother, and one from your father (that is, if everything goes according to plan). Since females have two X chromosomes, all eggs a woman produces get an X chromosome—because females don't have anything else to give. Males have an X and a Y, so half the sperm a man makes have an X and the other half have a Y.

Because of this, it is the *sperm* that determines a baby's sex: If a sperm with an X chromosome gets to the egg first, it's a girl. *X from Mom + X from Dad = girl* If a sperm with a Y chromosome gets there first, it will be a boy. *X from Mom + Y from Dad = boy* Now, you might already be starting to see why a man can't pass his baldness to his son. The baldness gene, again, is on the X chromosome. Dads don't give their sons an X chromosome; they give them the Y—that's what makes the baby a boy in the first place. This means that fathers can't give their sons any traits that are on the X chromosome.

Men inherit the baldness gene from the X chromosome that they get from their mothers. She may have gotten that bald-inclined X chromosome from her father, though, so an angry bald man could blame his shiny-headed grandfather if he really wanted to.

The baldness gene's address on the X chromosome also explains why baldness is more common in men than in women. And that in turn is the reason that baldness is said to be "sex-linked": you're more susceptible to it if your sex is dude—I mean, male.

Baldness is a *recessive* trait, and having a fully hairy scalp is *dominant*. What that means is that as long as you have at least one gene for nonbaldness, you're on your way to a full head of hair. You just need *one*.

The only way to go bald is to have nothing but baldness genes. Women have two X chromosomes, so they have two chances to get the nonbaldness gene. For a woman to go bald, *both* of her X chromosomes would have to have the baldness gene on it. The baldness gene isn't that common, so the odds of getting two X chromosomes both with the baldness gene are pretty small. Not impossible, by any means, but small enough to make a bald woman something of a novelty and for us to call the outcome "male pattern baldness."

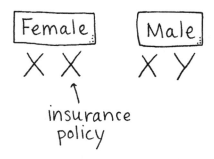

Ladies, if you do find yourself going bald, you have your mother *and* father to blame, because they genetically conspired against you to give you two X chromosomes, each of which had a gene for baldness lurking within its DNA. While I'm sure they didn't mean to do this, they nonetheless owe you dinner at a fancy restaurant. And don't skimp on the dessert, either.

In a man's case, to be bald he'd need one baldness gene on his X chromosome . . . and that's all. This is the real sticking point: men only have one X chromosome, so they only have one chance to get a nonbald gene. That Y chromosome of yours, boys, does not have any genes dealing with baldness. In fact, there really isn't a whole lot on the Y chromosome at all besides a gene for testosterone and other man-making hormones.

And so men are far more vulnerable to baldness than women because if that single X chromosome of theirs has the baldness gene, they don't have another X chromosome that can save them like the ladies do.

Now, if you're a man and you're wondering if you are someday going to go bald, look at your mom's family. If her father or brother is bald, you have a 50 percent chance of going bald too. But don't worry; you can use one of the many hair-growth products on the market. What you *can't* do is use them while thinking that you inherited baldness from your father. That would be silly.

The Ins and Outs of Blood Types

When I gave blood as a 17-year-old, everyone was freaking out.

"Be sure to drink lots of water beforehand."

"Don't drive there, because you definitely won't be able to drive back."

"Take your time getting up so you don't pass out."

I gave a pint of blood and felt absolutely normal, I swear to you. I was annoyed that everyone thought I was so frail and wimpy.

After I gave blood, the Red Cross sent me an official blood donor card. They would stamp it and write the dates of all my future blood donations. How thoughtful! And most importantly, it listed my blood type on it: B-. I got a mediocre grade on the blood test, it seemed.

What I didn't know at the time is that B- blood is exceedingly rare. Only two percent of the entire human population has this blood type. Turns out I'm not only hearty, but special as well. The blood bank would not stop calling me. It seems the only thing more rare than someone who has B- blood is someone who donates B- blood.

I went a second time a few months later, and after the pint of bloodletting, I'll admit I did feel slightly affected. I stayed in the lounging chair for about five extra minutes drinking apple juice and eating chewy chocolate chip cookies. But I was fine.

A few months later the calls started again. Won't you give us a few drops of your tasty red gold? Fine. That was the last time I donated. Because after resting for a good ten minutes, I got up to head to the restroom, and I unexpectedly had a run in with the floor. With my face. They continued to call after that, but I told them I had fifty tattoos, twenty piercings, and loads of unprotected sex, and so they left me alone from then on.

What was it about my blood that they adored so much? The proteins on my blood cells. When people ask you for your blood type, that's what they want to know.

A, B, O

There are three main varieties of antigens (a fancy word for protein) on your blood cells. There is A, B, and nothing, which we call O. A, B, and O.

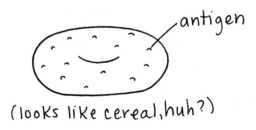

(looks like cereal, huh?)

Now, you have two genes for blood antigens—one from your mom and one from your dad. If you get two copies of A, you're blood type A. If you get one copy of A and one copy of O, you're still blood type A. The same goes for blood type B. Being B means you have either two copies of B, or one copy of B and one copy of zip (otherwise known as O).

If you get an A from Mom and a B from Dad (or the other way around), your blood type is AB. They both get expressed. None of that one-gene-covering-up-the-other business in this case. And lastly, if you get two copies of nothing, you are blood type O. It doesn't mean that anything is wrong with your blood cells; it just means that you don't express a certain protein on the outside of them.

Positives and Negatives

But I haven't even explained the plus and minus part of the blood type yet. That refers to another important surface protein on blood cells, which is called the Rh factor. Rh stands for rhesus monkey, because it was discovered in tests with the little monkey fellows. In this case, there are two possibilities: you have it or you don't. If you do have it, you are said to be Rh positive. If you don't have it (like me), you're Rh negative. But it doesn't mean that your blood is generally surly and miserable to be around. Far from it. My blood is the life of the party and tells lots of jokes and even stays after to help with the cleanup . . . and I think I've taken this musing too far.

Being Rh positive is dominant, because it only takes one Rh positive gene to make you Rh positive. You could get an Rh positive gene from your mother, and an Rh negative gene from your father, but you will still be Rh positive. It just takes one. You have to completely lack the positive gene to have a negative blood type.

Most people have the Rh factor, and are said to be positive. The gene that results in a negative Rh factor is on the rare side to begin with, so having two of them, and therefore having a negative blood type, is rarer still.

Matchy-Matchy

If you've ever wondered why O- is the "universal donor" and AB+ is the "universal acceptor" of blood, perhaps now you can see why. Let's discuss.

When a person needs a blood transfusion, that blood can't clash with the blood they already have. Most notably, the outside blood can't have any proteins on it that the recipient's blood doesn't already have. Those sorts of foreign surface antigens and proteins are what cause the body to say, "I don't recognize that! That's not us! Get rid of it!"

This is why O- blood, which basically doesn't have any of those proteins on it, can technically be given to anyone. It's totally bland and won't cause a scene.

For the opposite reason, someone with AB+ blood, who has all the antigens, couldn't possibly be given blood that has a protein the body won't recognize. But while anyone could take some O- blood and an AB+ person could take anyone's blood, hospitals generally aim to match you exactly to your blood type. I think they'd have to be extremely low on stock from a zombie apocalypse to start giving you blood that doesn't perfectly jive with your flow.

Another fun fact about matching: a pregnant woman and her developing baby can have some cohabitation issues if their Rh factors don't match—that is, if the mother is negative and the baby is positive. If the mother is negative, her body has never seen the positive antigen before—and if the baby has it, her body is going to wig out. It's called Rh incompatibility, or what I would call "pregnant lady blood rage."

More specifically, her immune system will start to wage an attack on the baby's blood cells. If this happens on a small scale, you get a jaundiced baby with yellow skin from all the bilirubin pigment that gets tossed around when the baby's red blood cells are destroyed. It's like the blood is bleeding. It's not good. If it's really bad, however, the baby might not survive.

As far as the ABO prevalence goes, most people are blood type O—about 45 percent of the population has it. Next comes A—40 percent of people have an A antigen on their blood cells. Then the Bs like me—11 percent. And lastly, having a gene for A and a gene for B, AB, accounts for just 4 percent of the population.

In terms of the pluses and minuses, it's the pluses that by far dominate. 84 percent of the population is positive (hooray positivity!). Negative blood types like myself account for just 16 percent. The only blood type rarer than B- like mine (which is 2 percent), is having both the A and B antigens *and* being Rh negative—that is just one percent of the population.

The B blood antigen seems to have originated in Asia thousands of years ago. Because it's a rare one, you can make maps of its prevalence across the world and see how ancient peoples' slow migration carried it westward. It's still far more common in Asia than anywhere else. This makes it particularly interesting for me, since most recently my ancestors are from Ireland and Scotland (hence my McName), where this blood type is exceedingly rare. I'm sure you're simply *fascinated* with my blood type and ancestry.

But now you can see why the Red Cross wouldn't leave me alone. The B blood type is uncommon and being Rh negative is also very rare, so with them together, I'm a magical unicorn of bloody desire. The only way I could be more enticing to them is if I was AB- or if I could give blood all the time without passing out on the floor.

Optometrists' Favorite Genes

Indeed, there is only one group that appreciates myopia, and that's optometrists. People like me keep them in business. Yet I would rather go to the dentist, get stitches, or break my thumb by slamming it in a car door than go to the optometrist. It's so traumatizing. They shoot air puffs at you, they put dye in your eye, and they dilate your retina so you can't be outside or see up close for hours. It's terrible. But it's possible that I feel that way because I have such a long history with optometry visits.

I got glasses in kindergarten. Yeah, I know. It was humiliating. I refused to wear them. I remember being in the car on the way to school with them for the first time. I was holding them in my hand, those blue-rimmed frames. I now realize that they were probably adorable. My mom gently suggested that I put them on. I put the arm of the glasses on my shirt collar, so they dangled from my neck.

She said, "No, on your head."

I put them on top of my head, the way I'd seen people store their sunglasses when indoors.

"Katie, put on your glasses."

So I did. I was a good little Katie.

But why did I need glasses so soon? And why is my vision to this day absolutely atrocious? A combination of genes and a few other factors.

Things that are uncool
in Kindergarten.

My parents both wear glasses, and all three of my siblings do now too (although none had to start in kindergarten). There is a very strong genetic component here, and dozens of genes have been implicated in causing nearsightedness. And yet, it's increasing at a rate that doesn't jive with the genetics. In the United States, nearly 40 percent of the population is myopic. In China and Taiwan, the incidence is up to 80 percent among children and young adults. What is happening?

The root cause of myopia has been the subject of a centuries-old debate. Is it genetics alone, or is it something we're doing as we develop to damage our eyes, such as reading a book inches from your face in dim light, staring at computer screens for endless hours a day, or perhaps sitting too close to the TV?

For the eye to process a sight, the light has to enter the eye and be directed onto the back of the eyeball at just the right angle, so that it will be in focus. It's like holding a magnifying glass up to a picture—if it's too far away, or too close, it doesn't fulfill its function. It needs to be at a specific distance to render the image the way it should. Myopia is like holding a magnifying glass a little too far from the picture. It makes things blurry. The technical term is "refractive error," which means the light misses its target, the retina on the back of your eyeball.

The genes that cause myopia belong to families of genes that control the growth and development of the eye, particularly the nerves in the retina, among other things.

If something isn't caused purely by genetics, we say it's due to "environment." This includes everything from the conditions in your mother's womb to your activities during development to the food you eat, the pollutants you're exposed to, and what movies you watch. Just everything.

The possible environmental causes of myopia are things like "near work," or anything that requires up-close focusing of the eye. A few centuries ago, near work would be quaint things like needlepoint or detailed calligraphy, no doubt by candlelight with harp music in the background. Today's near work consists of activities like reading books or looking at computer screens. But for Pete's sake, don't inform a struggling young reader about this. It could turn him off reading for a lifetime. What a choice that would be: Would you like to read wonderful books that open your mind to new ideas, or would you like to be able to see more than a foot in front of your face? Hmm. (I choose reading.)

But how is it that "near work" could be adversely affecting our vision balls? Because your eyes are operated by a series of muscles that could possibly be "misused" and cause problems later on.

Basically, sets of muscles in your eyes have to flex or relax at the right time for you to focus on a map of a hiking trail and soon after adjust to see a bear in the distance that has just spotted you. We do this all the time without thinking about it, but your eyes are busting their chops to achieve this for you. The worry about near work is that the muscles that are straining to focus up close then become less adept at seeing far away. I hear that some claim that eye exercises (staring off into the distance blankly, maybe?) can curb the effects of myopia and allow people to live glasses-free. Sounds pretty bogus to me.

Remember, bad eyesight is most definitely a highly heritable situation. If your parents both have glasses, the overwhelming likelihood is that you will need them, too. The uptick in myopia worldwide is attributable to the interplay between genetics and environment. For complicated traits that are the result of multiple genes, like eyesight, your DNA provides you with a likelihood that you'll exhibit something, but the choices you make factor into it, too. I was genetically destined to not have 20/20 vision, surely, but maybe if I hadn't done so much reading, holding books mere inches from my face, I wouldn't have quite the thick lenses I now require. I'll be sure to let six-year-old me know that when I buy my first time machine.

In any case, eye stress has always elicited highly opinionated advice. I remember being told not to cross my eyes because if I did it to excess, they would stay that way. Someone even went so far as to say that undoing it required surgery in which the eyeballs were removed and put back in facing forward. That sounded sufficiently horrifying to convince me to never give in to the temptation to cross them.

EYESTRAIN!

When Red and Green Look the Same

There are several types of colorblindness, but the most common is the red-green variety, which is what I will be focusing on here because I do what I want.

For these people, Christmas is a storm of confusion. Not just because of its overcommercialization and unsettling song lyrics (e.g., "Baby, It's Cold Outside [*and I'm a major creeper*]"), but because affected individuals' eyeballs can't tell the difference between red and green.

Red-green colorblindness is also called Daltonism, after the scientist John Dalton, who first described this situation. Colorblindness is caused by one measly little gene that is located on the X chromosome.

Living on the X chromosome makes a condition sex-linked, which you may remember means that your sex, male or female, is associated with whether or not you have a condition—in this case, colorblindness.

Colorblindness is recessive. As long as you have one copy of the gene that codes for full-color vision, you'll be okay. It's only if you have nothing but genes for colorblindness that you wind up colorblind. It's just like baldness that way. And just like baldness, it's more common in men than women. And for the same reason.

At the risk of sounding like a broken record, or corrupted mp3, or whatever it is we say these days: Women have two X chromosomes, and men only have one X chromosome; men's other sex chromosome is the Y. For a woman to be colorblind, she'd have to get a colorblindness gene from her mother and father, because she has two X chromosomes and therefore two chances to get a full-color vision gene.

On the flip side, a man would only need to get one gene for colorblindness because he only has one X chromosome. He doesn't have a chance to cover up a colorblindness gene.

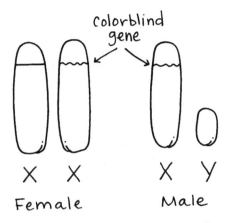

To understand how this one piddly gene affects something as complex as vision, we have to go into your eyeballs. But before that, let me take my contacts out.

Okay, I'm back.

Your eyeball. An orb of goo set in your face. Light enters through the pupil and is projected on the back of your eyeball, the part closest to the back of your head. It's not unlike a camera, actually. The back of your eye is your retina, and it is carpeted with photoreceptor cells that absolutely love getting light on them.

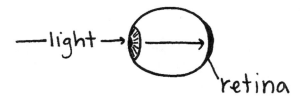

You have your rods, and you have your cones. The rods are really good at detecting light in general, as in "it's light outside" or "it's dark outside." That's really all they do. They're the more primitive side of the retina in that way.

The cones are the receptors that specialize in color detection. You have three different kinds—red, blue, and green. In this way, they're a lot like the old tube TVs. Do you remember getting really close to the TV and seeing the little red, blue, and green dots on the screen? It was so cool to see that all the colors of afternoon Nickelodeon shows were rendered in just those three colors. I loved it. Maybe this is why I am nearsighted. Hmm.

Back to that retina. Receptors do their job because DNA codes for proteins they need. You discern colors because cells in your retina have the specific proteins that can tell you "That there fire engine is red."

People with colorblindness got a bum gene for the pigment in their eyes that helps them detect the difference between red and green. The DNA itself has an error that causes a misshapen protein that in turn doesn't do its job.

Colorblindness, as far as genetic disorders go, is not so bad. Many people with colorblindness don't even realize they are affected by it. However, it can be quite a bummer for some when they find out there are whole careers they are not allowed to perform. Police officers, firefighters, and pilots are the most notable, and the ones children often aspire to be before they find out they are colorblind. Even less obvious jobs like meat inspectors, printer technicians, and casino dealers are at a disadvantage because of colorblindness and may be discriminated against.

For those colorblindees with dreams of being a firefighter, all hope is not lost. In the last decade, there has been some give on these incredibly strict rules, especially considering that colorblindness comes in different levels of severity. I sincerely hope that your eyeballs do not limit your vocational choices.

Those Baby Blues

I once went to the park with a toddler I was babysitting. She looked not entirely unlike me at the time. We both have brown, wavy hair; our skin color was similar, although she was a shade darker than me (which isn't saying much, since I'm so pale). A woman at the park saw us and struck up an awkward conversation with me.

"She isn't yours, huh?" she asked smugly.

"No, I'm her babysitter."

"Yeah, I can tell. Her eyes are brown, and yours are blue."

I gave her my, "Yeah, okay, whatever" face and went on with my babysitting duties, spotting little Hailey as she climbed ladders and went down slides. But what I should have said was, "You know, that is genetically possible, lady. So mind your own damn business."

The genetics of eye color have been billed as a straight-up Mendelian trait in the past with one gene with two variations—one for brown, and one for blue. Brown eyes were said to be dominant. If this were the case, two blue-eyed parents (who have two recessive alleles for blue eyes), could by no means have a brown-eyed child.

But they can. And it has nothing to do with the mailman.

Eye color is a tad more complicated than was previously thought. But the reason it was considered Mendelian for a time is that there is one gene on chromosome 15 called OCA2 that is very important when it comes to determining eye color. This gene controls the amount of melanin produced; one form of it causes brown eyes, and another form can lead to blue eyes.

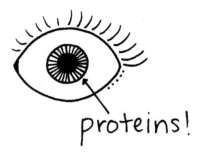

proteins!

But that's not the whole story by a long shot. There are a handful of, some estimate around a dozen, genes that are involved in determining the amount of pigment in your iris, and therefore what color your eyes are destined to be.

Our eyes have pigment just like our skin does. It has melanin, which includes two types called eumelanin (which is dark brown) and pheomelanin (which is orange-ish). If you have a lot of melanin in your eyes, much of it eumelanin, you have brown eyes, like that vast majority of people do.

If you have less melanin in your eyes, you may have blue, gray, or greenish eyes. Since no type of melanin is a blue pigment, it's not that your eyes *are* blue; it's that the light entering them bounces around, scatters, and bounces back out and appears blue to other people's eyes. Blueness is in the eye of the beholder, for real.

The way that light scatters and then registers to our eyes as "blue" is also the reason the sky appears azure to us. And similarly, there is nothing inherently blue about it—that's just how we see it. This is some seriously philosophical stuff here, man.

It's weird to remember that we're not truly seeing anything for what it inherently is, but rather we're seeing light bouncing off of it. We pick colors from it based on how our eyes receive those bounced waves of light and in turn relay the information to our brains. Our brains render everything and tell us, "That thingy is blue."

I hope you just looked around and realized that you're not seeing anything as it truly is and had a minor existential crisis. Those are fun.

Superseers

A strange thing is true of the mothers and daughters of colorblind men (and 50 percent of their sisters too). They have a superpower that allows them to see colors the rest of us can't. They are superseers.

Normal vision, as I described in a previous chapter, is based upon three different types of cones in our eyes. Each cone can differentiate between 100 different colors, so together they can perceive 10^3 colors, or one million. That's considered regular, average, Joe Shmoe vision. These individuals are called trichromats, because they have three functioning cones.

Colorblind fellas are called dichromats since they have but two working cones. They can only see 10^2 different shades of color, or 10,000. That's a lot less than a million. And I know that because ... math.

So what of these superseers? They have *four* different types of cones, so they are called tetrachromats, and this fourth cone allows them to see 100 times more colors than we lowly trichromats can. Superseers can distinguish between 100 million different colors. The extra 99 million colors they can see that the rest of us can't don't even have names, presumably since whoever thinks up names of colors is a trichromat.

How can you tell if someone is a superseer? One way is a test that asks you to observe three different computer-generated polka dots that to a trichromat look identical. Two of them are indeed exactly the same, but a third is just a slightly different shade that only a superseer could distinguish. An adept superseer will spot the dot that's a hair different every time.

Now let's figure out the genetic basis of these superseers. I mentioned before that this ability is seen in the mothers and daughters of colorblind men. That's how it was first discovered, actually.

In the 1940s, a Dutch scientist named H.L. De Vries was studying the vision of colorblind men with a test that challenged them to mix colors together to match a swatch they were given.

Just for funsies (or whatever word they used in the '40s instead of *funsies*), he did the test with the mothers and daughters of the colorblind men, and found that they seemed to distinguish *extra* colors. But he was a bit sheepish about this conclusion and didn't investigate further.

Why would mothers and daughters of colorblind men have this? Because they are carriers of the colorblind gene that in the aforementioned men caused their colorblindness.

Since the gene for colorblindness is on the X chromosome, women have insurance against it because being a woman means you have two X chromosomes. Having a functional gene on one X chromosome and a colorblind gene on the other X chromosome results in these women producing an extra, mutated cone. When this cone is produced in men, it puts them at a disadvantage in terms of vision, but these women—who have the genes for all three functional cones and this mutated gene for an extra one—get a special visual power.

What must it be like to have super color vision? I actually have no idea. But I wonder if, like supertasters, it's almost an annoyance. I wonder if superseers can detect mismatching colors that the rest of us think are identical, and get annoyed by it.

"Look! I managed to find a scarf that exactly matches these gloves!"

"Nope. Different colors. But don't worry about it, trichromat."

Also, I wonder if there are vocations that superseers would be especially skilled at, the way that certain careers are not advised for colorblind people. Creating and naming new paint colors might not be good, since the majority of paint shoppers are trichomatic and won't see a difference. I'm imagining a row of red paint swatches that all look distinct to a superseer and all look completely identical to me.

We know for sure that mothers and daughters of colorblind men are tetrachromatic, but exactly how many women are included in this superseer category? An estimated 12 percent of women are visually gifted this way.

I don't have broad knowledge of comic book characters, but this could be some serious science-inspired comic bookery. A superwoman who can see extra colors, and maybe also has some thermal or X-ray vision in there too to make it more fantastical. This would be, in my opinion, a great improvement to the other common superwoman powers that involve being invisible or having invisible forms of locomotion. Yeah, thanks for that one, comic book writers.

Straight, Curly, Brown, Blonde

Hair might be the part of our genetic selves that we change the most often. The hair cutting, dyeing, perming, and otherwise chemically or physically altering industry is booming, and it shows no signs of decline. People simply love messing with their hair.

The genes that determine the qualities of the hair you may want very badly to change are very complicated. You can tell just by looking at the crowd in line to buy groceries at Trader Joe's that there are a lot of variations. Brown and curly, black and straight, blonde and wavy, red and crazy curly; thick, thin, coarse, fine; it's a seemingly endless number of traits, variations, and combinations—and all of this just for some head protection. It's really weird, if you step back and think about it.

Whenever I need to take myself out of my own biased human experience, I imagine what it would be like to explain something to an alien.

"We humans have protein strands on our head."

"Why?"

"I don't know. To protect our brain or something. But it comes in all different colors, thicknesses, and textures. It's pretty neat."

"It seems like a waste of protein to me."

"Maybe it is, my little bald alien friend."

Have you ever noticed that aliens almost never have hair? Do movie producers and directors think that the presence of hair is an evolutionary blip that would happen only on earth and nowhere else?

Hair Type

Anyway, when it comes to determining the curliness or straightness of your hair, it's all about the shape of your hair follicle. In people with curly hair, the hair follicle has an asymmetrical shape that produces loopy doopy hair strands. Follicles that are bulb-shaped and symmetrical produce a very straight strand that doesn't have a curl to it. And in between those two extremes, you get the varieties of waviness.

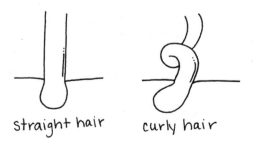

straight hair   curly hair

Hair follicles that are removed and grown in a lab continue to produce the same straight or curly hair that they did when they were on a person's head. So we've simplified the curliness of your hair to the hair follicle, but there are several genes that affect the shape of the follicle in the first place, and we don't completely understand them all. So one answer just raises more questions, as usual.

But we are starting to elucidate the functions of genes and what traits they contribute to. For instance, a gene called EDAR deals with baldness, the straightness of hair, and the amount of back hair you have. Yay, back hair! Who doesn't love back hair?

And another gene called PAX3 is associated with upper back hair, chin dimples, and unibrows. The fun never stops with those hair genes, I tell you.

Hair Color

Now on to questions of hair color. Much like your eyes, there are two main pigments that contribute to the hue of your hair: eumelanin, which is black, and pheomelanin, which is orange-ish. Red hair, as you can imagine, is found in people who produce more pheomelanin than eumelanin.

The vast majority of the population of earth (notice I'm not including aliens) has black hair, which is the result of lots of eumelanin production. Redheads, like I said, are biased toward pheomelanin, and blondes are too lazy to make much of either type of melanin.

The genes that cause the wide variety of hair colors are not fully delineated yet. We know some of them, but we don't have the full picture. But we get pieces here and there. For instance, a study of redheaded peeps in Iceland and the Netherlands found that the odds of having red hair were influenced by a gene called MC1R. In particular, they found that at a specific point in the gene, for each additional thymine in the DNA sequence, you are 6.1 times more likely to have red hair. This MC1R gene codes for a protein that is involved in making melanin, so that's why a change in it can affect your hair color.

And then of course there is the slow and steady march toward gray hairs, also called "aging," or something. You're doing it right now. So am I. I might be a few paces ahead of you, though, as I got my first gray hair at the ripe old age of 23.

A few years later, as an ancient 27-year-old, I discovered something quite curious. I noticed a gray hair on the top of my noggin, and although I've heard many a myth about the dangers of plucking gray hairs and the resulting funerals they throw, I threw caution to the wind and tugged on it. I inspected the silver strand, and noticed that my hair was gray at the end, but closest to the root, it was my natural brown. I didn't think this could be possible.

I mentioned this strange discovery on Twitter and got replies about how I must have gotten an incomplete dye job. I've never done such a thing.

I thought perhaps the gray stint was due to stress. I also considered the possibility that I was some kind of wizard or chosen one. So I searched the interwebs for someone who researches the metabolic processes that lead to gray hair. I found Dr. Desmond Tobin, a professor of Cell Biology and the Director of the Centre for Skin Sciences at the University of Bradforth in Great Britain.

He verified that hair follicles can, in fact, resume producing hair with pigment after a stint of gray, but that this happens in the early stages of a hair follicle going granny. So in other words, this particular hair follicle is trying really hard to not be old, but it's soon going to succumb to its gray-ness.

I also learned that as for blaming stress, there is actually no evidence that stress causes gray-ness at all. It is almost entirely genetically determined, with a small amount of environmental/dietary influence possible. So stress doesn't make you go gray, but being poisoned could.

So I'm not mad at you, hair follicle, for producing those gray hairs at such an early age. I'm mad at my DNA instead.

Magnificent Melanin

I'm going to go out on a limb here and say that you definitely have skin, as I'm rather certain that life would be impossible without it, and I don't recall ever coming across a disorder that rendered a person altogether skinless. But with biology, something that unusual is just a matter of time.

So, you have skin. What color that skin of yours is depends on the amounts of different pigments you have—if any. You might be albino. I have no way of knowing. We haven't met.

Pigments, by definition, are chemicals that soak up light. What part of light the pigment does not soak up, and therefore reflects, is what gives it its color.

Chlorophyll, for example—one of the favorite pigments of our botanical friends in the plant kingdom—doesn't appreciate the green part of light, so it gets reflected. That reflected green light is what hits our eyeballs and tells our brain, "Check it out. That plant be quite green."

I'm sure there's some deep philosophical significance to the fact that we only see plants as green because green light is what plants find the least useful and therefore let it bounce off of them. We see them for the color they actually hate most of all the colors. I guess it's sort of like getting to know someone by looking through his or her trash. Or judging someone's character by only talking to that person's ex-boy- or girlfriends.

And then, of course, there is the fact that there is nothing inherently "green" as we define that color about plants, just the actual size of the wavelength of light they reflect. It's our brains that interpret that wavelength and process it as the color green (which I consider my favorite color). A dog or a fly or an alien looks at a tree and sees it differently based on the receptors in their eyes and how their brain interprets the information. And furthermore, you are not truly seeing that tree; you're seeing the light that's bouncing off of it. So you're never seeing things for how they are in that exact moment, as the light has to have time to bounce off the tree and then bounce into your eyes. You are always nanoseconds behind the time. Nothing is real! Ahhhhhh!

Okay, back to melanin.

For a quick refresher about the "parts of light" I've been talking about, we have to discuss the electromagnetic spectrum. Whenever you read *electromagnetic spectrum*, try to hear a boxing announcer saying it into a microphone, the sound echoing off the walls of the stadium. It really helps with the overall effect.

The electromagnetic spectrum (ectrum, trum, um) covers all the waves that exist in the universe. Some, like radio waves, are all around us all the time, but we can't see them. Wavelengths that we can sense with our eyeballs are found in a very small portion of the whole continuum that we call the visible spectrum. It's the rainbow of colors that we see in the sky after it rains, the ROYGBIV: red, orange, yellow, green, blue, indigo, violet.

The wavelengths we see as red are the longest (and slowest) in this bunch. As we move to the blues and violets, the wavelengths are shorter and faster. If we keep moving up the electromagnetic spectrum, past violet is ultraviolet, often called UV. You've probably heard about avoiding that part of the spectrum, as UV rays can hurt us. Indeed, as the wavelengths get shorter and faster, they are more intense and full of energy. That is why blue flames are hotter than red flames. Keep your fingers away from them.

long waves are nice

short waves are mean

And now that we've begun talking about UV, we can talk about our happy little pigment that gives us a shield against it: melanin!

Melanin is a pigment that we produce in our skin (and our hair and irises). It gives our skin color, and it protects us from harmful light rays. As mentioned before, there are two kinds of melanin: eumelanin and pheomelanin. Again, eumelanin is dark brown and pheomelanin is reddish yellow.

The fair-skinned, red-haired, freckled individual like an Amy Adams or Prince Harry has more pheomelanin than eumelanin, and this is caused by a change in just one gene.

Apart from albinism, differences in skin coloration aren't the result of the presence or absence of different pigments you have, but how much of each one you make.

There's a type of cell in your skin called a melanocyte that produces melanin for you. If you have light skin (like moi), it's not that you have fewer melanocytes than someone with darker skin, but that your DNA instructs them to make less melanin. And having dark skin just means you make more melanin.

The great thing about melanin is that it absorbs and scatters ultraviolet rays that can hurt us. Those rays can penetrate our skin and scramble DNA within our cells, which is why excessive sun exposure can cause cancer: the DNA of tumor-suppressor genes gets damaged, and you're in trouble. We'll talk more about cancer in a few chapters.

Melanin is our shield and our protector, which is why the more melanin you have, the less likely you are to get skin cancer. And for this reason, sun protection is a hobby of mine. As I know from my Irish mother and grandmother, skin cancer is no fun *at all.*

You might be wondering why, when melanin is so clearly wonderful, anyone would have DNA that tells their skin not to make much of it. This is an evolutionary question: what would drive the natural selection of lighter skin? There must be some other force here. That force is vitamin D.

Our bodies use light for a few things, such as seeing (that's an obvious one), calibrating our brain's sleep-wake cycles, and making vitamin D. So while overexposure to the sun can cause cancer, living in a dark cave for a lifetime would really not be a viable option either.

If you have a lot of melanin in your skin, you need more sunlight exposure to make enough vitamin D, because you need some light waves to get past the melanin and get into your skin to the cells where vitamin D is being produced. When people started migrating to places where sunlight wasn't as plentiful, like parts of Europe, suddenly having a lot of melanin was presenting a problem. With little sunlight, and with the melanin soaking up so much of what little light there was, the body wasn't producing enough vitamin D.

Vitamin D deficiency can lead to rickets, a disease that causes bone softening and weakness. This is why milk ads are always touting its positive impact on bones—it has calcium (which is what makes bones hard) and vitamin D. But your bones will be just as happy with spinach and sunlight.

I know that my ancestors needed their vitamin D. I respect that. But I really resent them when I get a sunburn in twenty minutes flat. How is it possible that my ancestors evolved on this planet for millions of years and I can't go outside without the sun slapping me across the face and leaving a nasty red mark. How?!

With mild sun exposure, I do get tan. Never on purpose, mind you. I don't use tan as a verb. But there is a feedback loop in place here: exposure to light sends signals to my melanocytes to kick it up a notch and produce a bit more melanin to help me fend off those pesky rays. Some light-skinned people find this state of increased melanin expression appealing, and they roll around in the sun like a rotisserie chicken, or pay money to lounge in an electric skin cancer chamber (also called a tanning bed), to achieve the desired effect.

But at least I have *some* melanin. Albinos have absolutely no melanin at all. There are a few types of albinism, but one is called OCA1 for oculocutaneous albinism type 1. *Cutaneous* is Latin for skin, so there's that. This type of albinism is a mutation not in the gene for melanin, as you might expect, but instead in an enzyme that is necessary for producing melanin. As a result, there's no pigment in the skin, hair, iris, anything. The lack of eye pigment affects the vision of the person because there is also a lack of pigment in the back of the eye, the retina, which is what actually allows you to see.

Lefties and Righties

Being left-handed is in many ways a curse: you have to use the one subpar left-handed pair of scissors in grade school; when you write, you smear your handwriting as you drag your hand over the paper; and you bump elbows with right-handies at crowded dinner tables. It is rough. I don't know from personal experience, as I use my right hand for most things, but I am married to a lefty, his best friend is a lefty, and both his parents are as well. I'm positively surrounded!

In truth, the lifelong little stresses of being a lefty in a world designed by and for righties actually shortens the life expectancy of our left-handed brethren. And there is still discrimination against children who pick up a pencil with their left hand instead of their right. I thought this stopped happening generations ago, but I know firsthand (ha!) someone in my generation who in the late '80s picked up a pencil with his left hand and was told in no uncertain terms to knock it off. I have heard stories of this continuing. Who are these angry first-grade teachers?

10%. 90%.

Only 10 percent of the population is left-handed. An even smaller contingent—just one percent—are mixed-handed, or ambidextrous. They use their hands interchangeably. I had a friend in middle school who was obsessed with training herself to be mixed-handed because she wanted to be able to switch off between her hands when writing copious class notes so as to avoid hand fatigue. First-world nerd problems?

The genetics of traits that deal with your noggin are still a bit mysterious. Things that are the result of just one gene or traits that deal with purely physical traits—we're good at describing those, usually. But polygenic traits (ones informed by a truckload of genes) that concern the brain? Yikes.

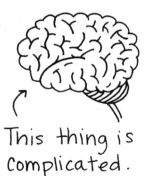

This thing is complicated.

Which hand you prefer is determined by how your brain is wired. There is a genetic component, but it's not entirely determined by what DNA you get from your parents. As far as we can tell so far, handedness is only 25 percent genetically determined. One of the things that could affect your preferred side are the accommodations in the uterus when the baby is developing. Stress has been implicated in the cause of left-handedness, which makes it sound negative, I know. But it's really not. Read on.

We know that handedness isn't completely genetic because identical twins aren't always like-handed. I can use my husband's family as a case study once again. My father-in-law happens to be an identical twin. He is a lefty, but his twin brother is not.

A gene that is likely involved in the handedness selection is a gene called LRRTM1. Yeah, it's a great name. This gene is involved in the pathways that affect the symmetry of the brain. This is also why genetic disorders like schizophrenia that are the result of asymmetries in the brain are more prevalent among lefties. But if you are a lefty, please do not worry about developing schizophrenia. There is simply an increased association.

On the other hand (ha!), being a lefty also makes you more likely to be creative and interesting. Think of just about any classical artist, scientist, or otherwise all-around cool person, and I can all but guarantee that he or she was a lefty: Albert Einstein, Jimi Hendrix, Marie Curie, Mark Twain, Helen Keller, Leonardo da Vinci—all lefties. Even Joan of Arc is said to have wielded her sword in her left hand. I don't know how we know that exactly, though.

A disproportionate number of U.S. presidents are lefties as well. And pay attention to scenes in movies where your favorite actor has to sign a document or write a check. You'll be surprised how many people that grace the silver screen are lefties as well.

I must say that I'm a bit jealous.

Genetics of Size

As we fight this "war" against obesity, we're on the lookout for genetic factors that make some people particularly at risk for being overweight. Sometimes I wonder if this information may have an unintended side effect—that maybe some people who are obese might not try to change their eating and exercising habits by claiming there is nothing to do, as obesity is fated by their genes. And the one thing genetics should not do is take away your hope.

Just like you might have a genetic likelihood to develop breast cancer, diabetes, or heart disease, if you know you have a genetic predisposition to be obese, that is not a reason to throw up your hands and say there's nothing you can do. Au contraire: It means that you have to work extra hard to stay healthy, the same way that someone who is prone to diabetes has to be careful around sugar. I know that doesn't make it sound fun, and may seem "unfair," but in genetics there is no fair. There is the DNA you're given and what you do with it. We all have our DNA annoyances.

Recent studies show that obesity is between 64 and 84 percent heritable, meaning that having obese parents does indeed make you more likely to be obese yourself. Of course, it's an estimate for now because teasing apart the genetic factors and environmental influences in this case is extremely difficult. Are children of obese parents more likely to be overweight or obese because they are eating the same food and have the same level of activity as their parents? There are a lot of factors to consider.

We haven't tracked down the exact list of genes that can lead to obesity, but one that looks particularly guilty is called the FTO gene. Certain forms of this gene are correlated with an increase in weight.

The FTO gene codes for an enzyme that modifies DNA, and it seems that in fatty tissues this gene is more active. Looking mighty suspicious there, FTO. Tsk, tsk, tsk.

Besides genetics, diet, and exercise, the particular ecosystem of gut bacteria you have can influence how much weight you do or don't gain from your diet. And what bacteria you have roaming around in there is greatly influenced by what you eat every day. You shape the colonies with every meal you sit down to.

When it comes to the other part of the size of your body, your height, it is about 80 percent genetically determined. The other 20 percent falls under that "environment" umbrella, which can include things like what you eat, what you do, and what your favorite color is (no, not really). A healthy diet while you're growing is, not surprisingly, important for reaching your full height potential.

One of the genes involved in height determination is named GDF5. This gene has instructions for a protein that serves as a growth factor in bone. But strangely, this gene only seems to affect height outcomes in people of European descent. I have absolutely no idea why this would only be true in one group of people. Perhaps the mutations in GDF5 that lead to the differences just haven't happened in other groups. Genetics is really weird that way.

We're not done yet on the subject of tallness and shortness. In the next chapter we'll look at the possible extremes when it comes to how tall you are.

Dwarfs and Giants

People come in a pretty amazing variety of sizes. I am five feet, eight inches (1.73 meters) tall, which happens to be the height of the average man, which is also the height of Mark Wahlberg. If I ever meet that guy, I'll be sure to let him know that we are exactly the same height and will suggest we touch noses to prove it. Yes, these are the things that occupy my brain.

The people vying for the "shortest person in the world" title were or are twenty-something inches (0.5 meters) tall. The tallest people ever have been nearly nine feet (2.7 meters) tall.

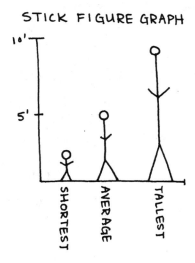

STICK FIGURE GRAPH

The people at either end of the spectrum, of course, have certain genetic conditions that cause this. Something as complex as a person's height has many opportunities to go a little haywire, but here I'll just focus on two particular conditions: dwarfism and gigantism.

Dwarfism isn't caused by one particular genetic avenue; there are several ways it can happen. Under the dwarfism umbrella are a few medical conditions that cause it. One of them is called achondroplasia. It's caused by a mutation in a single gene. The reason it has such a broad effect is that it's a change in a gene that is involved in development. The most common change is one amino acid in a protein found in cell membranes.

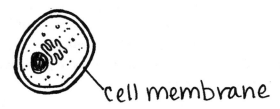

cell membrane

The good news is that while this causes shortening of limbs and a slightly larger noggin, it doesn't usually have any other serious effects. Life expectancy and brain function are completely normal. Rejoice!

Because achondroplasia and other causes of dwarfism are a dominant trait, dwarfs can be carriers for the gene that leads to regular (or shall I say, average) height. That means that two people with dwarfism could have a child who is five foot ten (1.78 meters). If you have watched *Little People, Big World*, or even seen an ad for it, you may have noticed that the parents have two children that are dwarfs, and two children that are not. Parents Matt and Amy Roloff have different forms of dwarfism, I should mention. Amy has achondroplasia, but Matt has a rarer form of dwarfism called diastrophic dysplasia.

The opposite genetic treat is acromegaly, sometimes called gigantism, which many people associate with André the Giant, André René Roussimoff (from *The Princess Bride*).

Acromegaly is caused by the pituitary gland going a bit haywire often due to a tumor. This causes the overproduction of growth hormone, which does just what it sounds like it does. This is tricky because it often happens so gradually that it doesn't seem to be a problem until it has progressed significantly. And it can get dangerous fast at that point. We're talking osteoporosis, diabetes, and even heart failure.

So what do you do if you start developing gigantism? Get rid of that crazy tumor that's causing the ruckus. Surgery, radiation, and hormone therapy. No fun, that's for sure. But acromegaly is dangerous.

As for the exact genetic cause of acromegaly, well, that's still being worked out.

Researchers have been testing the DNA of extended families with a higher occurrence of gigantism to find the exact gene that is causing it all. Since it's due to a tumor, it will most likely be a mutation in a gene that would otherwise protect the body from cancers in the pituitary gland. How could a gene go about doing that? That just so happens to be what the next chapter is all about. Read on.

Stupid Cancer Genes

Cancer is quite the counterintuitive situation. A cell goes against all the rules and starts dividing uncontrollably. It multiplies and spreads, splitting in half again and again. This reckless cell growth produces a tumor, which can compete for space with other— normal, healthy—cells and disrupt organ function. Worse yet, these cancerous, madly multiplying cells can hop aboard the body's transit systems and spread to other parts of the body.

Why on earth would the body do this to itself? It makes absolutely no logical sense. Are our bodies self-hating masochists?

Cancer comes seemingly out of nowhere, and it's hard to treat because it isn't a foreign body like a virus or bacteria that can be isolated due to its differences. No, cancer is *you*, so targeting it is incredibly difficult, and the treatments we have, like radiation and chemotherapy, are grueling because they don't just hurt the cancer, they hurt you too. In short: cancer is the worst.

I've heard people blame cancer on our modern lifestyle, with our post-industrial exposure to pollutants, a diet rich in processed foods, and our at-screen-staring while on-butt-sitting tendencies.

But wait a minute there, you. Cancer is not specifically a disease of the twentieth century and beyond. It's been around for quite a while. Ancient Egyptians had cancer. Early hominids had cancer. Even dinosaurs had cancer.

It's been around for a very, very long time. And it's not going anywhere.

How does this happen? How all of a sudden does a cell or group of cells decide to ignore everything it's ever been told and start multiplying and spreading and damaging your organs? Mutations. Mutations in the genes that normally tell cells to avoid doing those idiotic things.

A gene that when functioning correctly keeps cell growth in check becomes a cancer gene when it has a mutation that keeps it from doing its vital job.

The most talked-about cancer gene is the so-called "breast cancer gene." This is misleading, because we all have this particular gene, but certain mutations of it can cause cancer. There are two known genes with a link to breast cancer: one called BRCA1, which sits on chromosome 17, and another called BRCA2 that's on chromosome 13. If you have a certain form of one of these genes, the risk of breast cancer jumps from 13 percent to 60 percent.

These BRCA guys belong to a family of genes that are called tumor suppressors. They code for proteins with a variety of functions like repairing damaged DNA and controlling crazy cell growth. This is precisely why if you have a tumor suppressor gene with a slight problem, you're far more susceptible to cancers. Those genes and their products aren't quite right, and they can't fulfill their cancer-preventing destinies.

And it's not just breast cancer. There are hundreds of genes that are linked with cancers ranging from skin cancer to colon cancer.

So what are we to do? Well, if you're a big fan of information, you can get your genome sequenced and see if you have one of the known cancer genes and find out what your risk is for developing a certain type of cancer. And then, if your name is Angelina Jolie, you can surgically remove the area that is threatened. I do know of other people besides movie stars who have elected to have this surgery after watching all older female family members die of breast cancer. Breasts aren't kidneys. You can live without them. And plus, bras are expensive.

A lot of people criticized Angelina Jolie's decision to have surgery. They question, where does it end? Should we just surgically remove body parts in order to decrease our risk of cancer? Others balked that removing her breasts doesn't decrease her risk to zero, so it was a waste of time. Furthermore, sequencing your DNA is not cheap, although it is getting cheaper as we speak; so even if you're not a multimillionaire, this can be easily within financial reach. (I'll talk more about this later.)

To all the haters and naysayers and negative Nancies, I say this: What do you care? Cancer is a disease that is personal to each individual, not least because it's their own body that's causing it, not something like the flu virus we all were exposed to. Your cancer is different than my cancer is different than Angelina's cancer. How people think about it, respond to it, and plan for it is as personal as the cells that cause it. So calm down. Sheesh.

Troublesome Genes

Genetics isn't all fun and games and pea plants. There are very serious, very debilitating genetic disorders out there, and plenty of them are caused by just one faulty gene. That is the thing about life—it takes a million moving pieces for it to work right, and just one to mess everything up.

There are more than 10,000 single-gene disorders, most of them incredibly rare. Less than 2 percent of the population has any of them. But they're out there—disorders like cystic fibrosis, Duchenne muscular dystrophy, hemophilia.

Genetic disorders are generally rare because individuals that have them may not pass those genes to the next generation by having children. Especially, and obviously, if the genetic disorder is fatal early in life, before an individual would even have the opportunity to have kids.

Seeing as how it was only very recently, in terms of human evolution, that the real pressure of natural selection was lifted from us—with the advent of medicine, flushing toilets, and soap—you'd think that there would be almost no genetic disorders that would remain in the populations of our ancestors. You'd think they would be weeded out pretty fast.

Of course, they often are, but that leaves out a few considerations:

1. If the genetic disorder doesn't cause much of a problem until you're older (and have already had kids), it will keep getting passed on.

2. This doesn't account for recent mutations that can simply just come out of nowhere due to chance.

3. Sometimes there is a slight benefit associated with a genetic disorder.

Most of the diseases associated with old age, like Alzheimer's, are not subject to any kind of natural selection because they happen after you have kids and pass on the genes that contribute to that condition. If Alzheimer's set in during adolescence, it probably wouldn't have remained in a population of people for very long, as being a parent would be, shall we say, severely hindered if you couldn't remember anything. So that's one way that genetic disorders are able to stick around: you don't have the opportunity to select against them, since they're sleeper cells.

Genetic disorders can also crop up in a population of people because this crap just happens. It doesn't necessarily have to be inherited from a parent. Trisomy 21 (Down syndrome), for instance, doesn't occur because parents have it or are carriers for it. It just happens because chromosomes don't separate correctly when the sperm and egg are forming. There's no way for that to be selected against if it just happens sometimes.

The third example is undoubtedly the one that made you furrow your brows into a "What?" face, so let's go there.

Sickle-cell disease is caused by a single base pair mutation in the gene for beta-globin, one of the pieces of hemoglobin, which is an important ingredient in your red blood cells. What this means is that if you have this error, you have blood cells that are a tad misshapen. Rather than the puffy disc shape red blood cells should have, the blood cells of someone with sickle-cell disease are tapered and C-shaped. They look like a sickle, which is why it's called sickle-cell disease. Science!

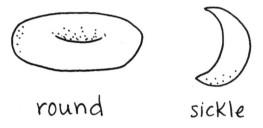

round   sickle

People with sickle-cell disease deal with the fallout of having these oddly shaped blood cells. For starters, the misshapen blood cells are frequently destroyed by the body, which leads to severe anemia. Red blood cells' job (well, one of them) is to carry oxygen to all your tissues. If you're low on red blood cells, your body is going to be low on oxygen. Being anemic makes a person feel weak and horrible. It's no fun at all. And if your heart isn't getting enough oxygen, that's the worst, as it spells possible heart failure.

And on top of that, the sickle shape of these blood cells makes them far more likely to cause a blockage in blood vessels. The pointy curved shape of the sickled cells are almost too perfect at forming clumps, since they can stack so easily. Blockages in blood vessels are a very serious situation: they can cause extreme pain, kidney failure, and strokes. The round, pool-floaty shape of regular red blood cells don't really allow for this.

So why am I using this as an example of a genetic disorder that has a good side effect? It sounds incredibly troublesome so far. But what I just described is what happens if you get *two* copies of the gene that causes this condition. If you get just one copy of it, and have a functional beta-globin gene as well, it gives you a very special power: increased resistance to malaria.

Malaria is caused by a protist called *Plasmodium falciparum* that hops aboard a mosquito and winds up in our blood when that mosquito engages in the age-old pastime of drinking our blood. People who have one copy of the gene that causes sickle-cell disease have an advantage because when the parasite invades the red blood cells, they change shape and are quickly destroyed by the body.

An individual who doesn't have any genes that cause sickle-cell disease doesn't have this protection, so malaria can take its typical course, which is flulike symptoms at the start and convulsions and coma at the end. No, thank you.

So even though having two copies of the gene for sickle-cell is most definitely problematic, having just one can be extremely helpful if you live in a part of the world where malaria is common. And for this reason, sickle-cell has been conserved in some populations. But the obvious downside is that two carriers who get the benefit of the resistance to malaria can potentially have a child that gets both their copies for sickle-cell and winds up very sick indeed. It's a natural selection tradeoff, so to speak.

Other genetic disorders may have increased resistance to other diseases in our ancestors' past. For instance, it's possible that Tay-Sachs disease gave Eastern European Jews a slight resistance to tuberculosis. Cystic fibrosis may have lent a hand to Europeans dealing with the plague and cholera. And type 2 diabetes may have come in handy when dealing with periodic starvation. Good times.

Five Fingers?

I may have said that there aren't many complex traits that are due to just one gene. But the same is not true of *mutations* to those genes. Change one little gene, and it can have a serious effect. If that one gene happens to be involved in development, you're talking some pretty dramatic changes.

When you were but a tiny fetus (and looked like a bean), there were genes at work that were helping to set up how your body was going to form.

Ha!
Look at you!

Some genes make proteins that act as a "brain goes here" or "butt goes here" sign. And from that first decision to make one end of you a head and the other end an ass, you started to develop. Later, decisions about where to put arm nubs and where to have fingers and where to not have fingers were made. If these genes aren't completely accurate, you can get some deviations.

One such case is polydactyly, or the state of having extra fingers and/or toes. Believe it or not, this is very easy to do with just one gene. And this gene is also dominant, which might sound surprising. But being dominant doesn't mean a gene is common. This particular dominant gene is exceedingly rare. What's also interesting about polydactyly is that there isn't just one gene and one mutation that can bring it about. There are several different ways for it to happen. And sometimes polydactyly is just a side effect of a different genetic disorder.

When your hands formed in utero, they first looked like goofy hand paddles. It's only when cells in what should be spaces between your digits die off that you start to get the familiar chubby baby digits we all expect to see. Because this is happening in so many areas and must happen at a specific time, there are several ways this can go awry.

Luckily for those that have this mutation, surgery can often solve the issue immediately. While there wouldn't be anything really wrong with having a sixth digit on your hand that functions, often polydactyly can result in a partial finger that interferes with its neighbor. And that simply will not do.

not super helpful

But something to consider when dealing with a dominant trait like this is that corrective surgery does not erase your chances of passing it on to your child. If you have one gene for polydactyly and one gene for regular hands and feet, that means you have a 50 percent chance of passing it to any children.

In that way, it could become a bit like bad eyesight. The genes for bad eyesight in ancient peoples were probably weeded out, as the actual survival of a person with blurry vision was less likely. Personally, I probably would have fallen off a cliff or been eaten by a bear if I had been born before the age of civilization and soft contact lenses.

Since I can correct my vision with glasses, contacts, or Lasik, I can get by just fine, but I will undoubtedly pass my less than stellar eye vision genes on to any children I might have. And so have a lot of other people, which is why having bad vision is in some places more common than having 20/20.

Bubble Boys to Superheroes

Bubble Boy isn't just a weird movie with Jake Gyllenhaal and Danny Trejo. It's also the depiction of a real disease. The actual name for the condition is severe combined immunodeficiency, or SCID. Indeed, it is severe and combined and also immunodeficient.

The Bubble Boy term and aforementioned movie was inspired by a real person. David Vetter was born with SCID in 1971 and immediately placed inside a protective plastic environment called "the isolator" until a bone marrow transplant could be performed. When the transplant failed to cure him, he stayed in the bubble. He died when he was just twelve years old.

Today we know a great deal more about this condition, and no one will ever again need to live inside a protective bubble—unless that's their thing, in which case, no judgment.

Like most things that are complex, such as the whole immune system, there is a big family of genes that contribute to it. But that doesn't mean that one gene can't royally screw everything up. Ah, life. It takes enormous and constant effort to build life's skyscrapers, and just one giant lizard to ruin everything. It's just not fair.

In this case, one mixed-up gene can ruin the whole immune system operation because of the way your immune cells are made.

The immune cells that fend off invaders like viruses, bacteria, and fungi are made in your bone marrow by a cascade of processes. If a protein at the start of that assembly line has an error, everything downstream is nonfunctional. That's exactly the case with SCID. There are actually several different genes that, with a mutation, cause a form of SCID, but they all involve changes to proteins that are involved in making immune cells.

Babies born with SCID that goes untreated don't usually survive for more than a year, as they die from things like pneumonia or chicken pox. But today SCID is not a death sentence or live-your-life-in-a-bubble sentence. With a bone marrow transplant, SCID can be cured. Let there be much rejoicing.

Now that we've talked about how horrible it is to have a completely nonfunctional immune system, let's consider how we strive to give our offspring the best immune system that money can buy. Or that DNA can buy. You know what I mean.

On the other side of this immune system genetics coin are genes that are specifically aimed at improving the immune systems of your babies. Even your choice of a mate reflects your instinctual urge to produce children with rockin' immunity.

You have a gaggle of genes that code for something called the major histocompatibility complex, or MHC. MHC molecules play a big role in your immune system. They're expressed on the surface of cells and are involved in how your immune system determines what falls under the umbrella of "you" and what falls under the other umbrella of "No, this is definitely not you. Get rid of it immediately!"

Having top-notch MHC genes is key to having an effective immune system. And to have babies that have A+ immune systems, it's best to find a mate whose MHC is most dissimilar from yours. In genetics, just like in investments, diversity is a fabulous thing.

But obviously you don't know what your MHC looks like, and neither do the people you date, so how on earth can you choose a mate that has a different and therefore complementary MHC to yours?

You use your schnoz! Seriously.

Your dating coach

hehe... boogers

This part is almost so cool that it's creepy. People who have an MHC that meshes nicely with yours *smell better to you*. Yes. You think that Mr. Man smells awesome? That's because your hypothetical babies would be healthy. Yeah. This biology thing knows you way better than you know yourself. You're not picking your mates; your secret nasal preference for immune complexes is. Who am I? Who are you? Who are we? What is happening?!

I don't know, but just calm down. There is obviously more to a person than how they smell. This is just one piece of the puzzle. But it does give some perspective about our bodies, our life choices, and our pursuit of mates. There are biological mechanisms in place that help guide us through these things. And it's a reminder that whether we like it or not in our modern existence, our evolutionary primary objective is to reproduce. You might think it's just to survive, but that's not the whole picture. Surviving is a secondary goal. It's really just a means to reproducing.

But this purely evolutionary goal is not necessarily your life goal. We are not actively engaging in evolution like our wild relatives are, where survival and reproduction are really just for the fittest, as the others die of starvation, from disease, or in the mouth of a predator. It's rough out there. We sort of stepped off the evolutionary treadmill when we developed medicine and computers and McDonald's hamburgers. But it's still in the background, affecting our decisions. Like a ghost haunting us.

Now let's talk about another spooky facet of the genetics of immunity: viruses and the souvenirs they leave behind.

Viruses love injecting their DNA (or RNA in some cases) into our cells in order to make more of themselves. That's how they cause such a ruckus. They hijack our poor cells and force them to become virus factories. It's awful. But sometimes those viruses inject their DNA and it just sits dormant. The viral DNA gets incorporated into our DNA then, and when our cells replicate DNA before dividing in half, the viral DNA gets copied too.

Because of this about 8 percent of our genome is the leftovers of generations-old viral infections. In this way, we can see what infectious diseases our ancestors had to deal with.

They're viral souvenirs. Sort of like a keychain that gets embedded in your body. Yay?

Allergy Fun Time

Inhalation allergies. A chronic and maddening condition in which our immune system overreacts to harmless proteins, freaking out needlessly, launching an all-out attack on the substance, and in the process, making us simply miserable.

Interestingly, the symptoms of the common cold are due to a similar overreaction of the body to the rhinovirus, the virus that causes this well-known malady.

The rhinovirus...

looks like a Koosh ball.

The runny nose, sore throat, and general sense of ickiness are not caused by the virus itself, but by the body's inflammatory response and attempts to flush the virus out of the system. Especially the snot. It flows freely to try to wash the offending pathogen from your upper respiratory system. This might work, but the snot also slides down the back of your throat and triggers pain receptors there. In general, these alarm bells do very little to actually defeat the common cold, but on the plus side, they do allow you to call in sick to work.

Allergies in general are a problem of the overly clean first world. Our ancient ancestors probably rarely saw anything that resembled allergies. Their immune systems had real problems to deal with, so their defenses were calibrated "correctly." They prioritized their powers for real threats of disease and parasites, and simply had no time to fuss around with pollen and dander.

One in five people in the United States report problems with allergies. I'm one of them. I have mild allergies to cats and dogs, and in an allergy test a few years ago, they said Bermuda grass as well. I do have memories of rolling down grass hills as a kid and ending up with welts all over my bare little legs. But doesn't everyone get that? Or is that just one of those things I assume is universal and is really just me?

Allergies are still virtually unknown among children who grow up around a great deal of dirt, such as on a farm. Getting dirty is one of the best ways to avoid allergies later in life (and being breastfed gives you a head start too). Exposure to all the would-be allergens in dirt, dust, and other such filthy fun can familiarize your immune system with these harmless neighbors and avoid the later freakouts.

Allergies are right now more common in women than men, and it might be because of the cultural tendency we have to dress little girls up in nice outfits and urge them not to get dirty. If little girls were just as encouraged as boys to go outside and play in the dirt and eat bugs, there would likely be a similar incidence of allergies between girls and boys, women and men. I never ate any bugs, but I did try some dirt in preschool. I didn't really care for it.

Food Allergies

So far I've just been yammering on about the inhalation sort of allergies. Those are definitely annoying, but are nowhere near as serious as the food allergies. Those can kill you.

The allergic response to food baffles me. It is probably the least adaptive thing our body does. Normally, our bodies are adept at filtering out toxins, regulating all our systems, and prioritizing activities—like how your digestive system shuts down when you're stressed so the resources can be earmarked for muscles such that you can escape an impending bear attack (or public speaking engagement).

But when the body decides to shut systems down and launch a massive inflammatory response because of something you ate, it serves almost no purpose. It doesn't rid you of the offending food. It just causes misery and can possibly murder you. Why do we do this?!

Loads of research is currently under way to unravel the causes of different types of allergies. Food allergies are caused by a mix of genetic and environmental factors, just like most of our complicated traits. The recent increased prevalence of food allergies suggests that there is more than just genetics at work here. It's possible there is an *epigenetic* effect at work here as well.

I'll go further into epigenetics later, but basically: epigenetics describes how our DNA can be changed not by changing the actual sequence, but by changing which genes we do and don't use. Often we describe this as genes being turned "on" and "off" by means of a switch.

I'll give you an example. As you have surely heard, it's a bad idea to smoke while pregnant. It's also a bad idea to smoke any time, but I'll not lecture you on that. Everyone already knows (and yet people still pick up the habit. Hmm.). Babies of mothers who smoked while pregnant are more likely to have asthma later in life. The mother didn't pass the genes for asthma to her child; through her actions, genes were turned on and off that resulted in asthma. As Charles Darwin surely would have said, "Yikes."

More chillingly, if that developing baby is a girl, and therefore is simultaneously forming eggs inside her fetal self as she develops, the grandchildren of a mother who smokes while pregnant could have asthma because the smoking affects the DNA in the eggs of the new baby. This epigenetic stuff is not messing around.

The genetic causes of allergies are going to take some time to be completely understood, but in the meantime there are loads of pseudoscience to fill the void. Be very careful where Google takes you on this subject. Most of it is conjecture masquerading as science.

People have suggested that GMO foods are to blame, that plastics are to blame, that Obama is to blame, etc. While everyone fights about which single cause is responsible for a complicated problem, researchers will probably quietly discover that the contributing factors are many, interrelated, and varied. That's usually how this goes.

Medically Necessary Gluten-Free

Food allergies seem more prevalent than they really are because awareness and marketing for them has increased manyfold. One such allergy is due to celiac disease, which renders a person unable to process a protein called gluten that is found in wheat products (and barley and rye, which are not as common as wheat, unless you're talking about beer).

These people have to eat gluten-free food (and not drink beer), which of late has become quite the trend. I see the evidence frequently in food labels: foods aren't just "low-fat," they are also proudly "gluten-free!"

But don't let the fashionable gluten-free diets fool you. Celiac disease is a very rare condition with an incidence of one percent, and, interestingly, it's far more common in women. But for those that have it, it's legitimately no fun. Their immune systems simply can't stand gluten, and when it makes its way to the intestines, the body launches an attack not just on the gluten, but on the intestines themselves. It's incredibly misguided. Seriously, immune system. You really need to get your crap together (the way the intestines do. Hehe).

The genetics of gluten frustration is not fully understood (and I bet you're getting tired of my prefacing everything that way!), but one way it can occur is with a mutated form of a gene called HLA-DQ. No, it has nothing to do with Dairy Queen.

HLA-Dairy Queen has instructions for making a vital protein in the immune system that attaches to suspicious proteins and acts as a target that says, "Attack here, guys." This altered form of HLA-Dairy Queen has a special fondness for gluten. For no reason whatsoever. What did gluten ever do to you, HLA-Dairy Queen? Leave it alone! Gawd!

What's especially weird about the HLA-Dairy Queen is that this gluten-hating version of the gene is far more common than full-blown celiac disease. Approximately 30 to 40 percent of the population has this gene, but only one percent (like I said earlier) actually can't put up with gluten. It's possible that the immune system's gluten rage is just waiting for first contact with gluten outside of the intestine. Following an intestinal injury where gluten gets below the surface of the small intestine, the immune system would be alerted to its presence and then be more sensitive to it later.

Celiac disease is not to be confused with gluten intolerance. This is not a full allergic response to gluten, but more of an intense dislike or annoyance. Many people report feeling overall better if they avoid gluten altogether, and feeling lethargic after a bread binge (mmmm ... bread ... drool), but that does not mean they are allergic to it.

People sometimes use *allergy* and *intolerance* interchangeably, and that really annoys me.

Here's a good example. Many people are lactose intolerant, which means that past the point of breastfeeding, around one or two years old, they stop producing an enzyme called lactase, which is in charge of breaking down the sugar, lactose, that is found in milk. If you don't make lactase, then partaking in dairy products makes you farty, uncomfortable, and stinky.

A dairy *allergy* is something else entirely—as with other food allergies, it causes throats to swell and hives to appear. That is not lactose intolerance; that is lactose raging murderous immune system.

I look forward to a day when new medical treatments will make it possible to reason with our overactive immune systems and say, "Calm down, you silly bastard. It's pollen. It's a peanut. It's gluten. For the love of science and all that is rational in this world, keep it together."

And if you have children now or will soon, make sure they're absolutely covered in dirt, dust, pollen, and dander perhaps 90 percent of the time. Do it for their future.

Mysteries of Your Mind

Depression, anxiety, schizophrenia, bipolar disorder, attention deficit-hyperactivity disorder. It's a jungle in there, our minds. As I've said about a bajillion times already, any complex system that requires a bazillion things working together to function has just as many ways to go a bit haywire. Nowhere is this more obvious than in the cavernous, mysterious Jell-O mold that is the brain.

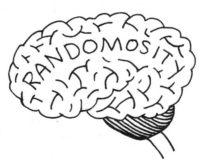

Mental health has become a bit of a buzz term lately. It's frequently discussed because the U.S. health care system often is underequipped to deal with it, and in many cultures, mental disorders are still seen not as a physical or chemical malady, but as a sign of weakness that must simply be "gotten over," like a stubbed toe.

Mental health is yet another area where we are getting the fuzziest of pictures about how genetics and environment (nature and nurture) come together to bring about these mental health problems.

Here are some of the things we do know. Depression is 40 percent heritable. Schizophrenia is highly heritable if you group full-blown schizophrenia and schizoid personality disorder together. In this case, if one parent is schizophrenic, there is a 49 percent chance the child will be as well. If both parents are, the incidence jumps to 66 percent. If one identical twin has schizophrenia, there's an 87 percent chance the matching twin has it too.

But remember that these conditions aren't all-or-nothing states like wet or dry earwax (remember that?); they exist as a broad continuum. So when I say that depression is about 40 percent heritable, it doesn't mean 40 percent of your children will definitely be depressed. It means that there is a 40 percent chance a child of depressed parents will be susceptible to developing depression as well. So it is by no means a definite outcome, just a general forecast, many times predicting a storm that never shows up on the horizon.

In the continuing search for the genes responsible for mental illness, there are four DNA snippets that are likely, and two of them deal with calcium channels in cells. The way that your brain sends the electrical signals to convey something like "This burrito is my best friend" is by moving charged chemicals (ions) around to create positive and negative ends, like a battery.

Calcium ions are positively charged, and if the calcium channels aren't working right, this can change how nerve cells communicate.

And interestingly, three of the four possible DNA mutations were associated with multiple mental illnesses: autism spectrum disorder, bipolar disorder, ADHD, major depression disorder, *and* schizophrenia. This phenomenon was foretold in twin studies in which identical twins had different mental illnesses. Dun dun *dunnnn.*

Mental illness awareness-raising is a tough job because you don't want to send the wrong message, and it's so easy to. You don't want to overdiagnose conditions like ADHD and depression, nor overmedicate.

There is also a strange sense of blame when it comes to mental illness. Some, when they find out their problem is due to a genetic, physical problem that is not the result of their failing in any way, are ecstatic, as it means that they have done nothing wrong and can now search for a solution without guilt.

Others, though, feel doomed by the physical and genetic nature of their illness, as it means that it's a concrete problem that is not likely to magically go away. But with support, and in some cases medication, there is a lot that we can do to help. And as we gain more knowledge about the nature of these mental health problems, the options for care will only increase.

Your You-ness

This is the information that most people want to see unraveled: how do our genes affect who we are as individuals? What genes contribute to high IQ or musicality or tennis-playing skills? What section of DNA makes one person stay positive in the face of extremely difficult circumstances and another person a cynical slug despite having advantages and opportunities that few people ever know? How do our genes contribute to that general essence each one of us thinks of as our *self*? The answer is pretty straightforward: We don't know yet. I'm sorry to be a disappointment, but it's true.

The things we qualify as our personality, the summation of traits that make us unique, all reside in our neurobiology—the way our brains work. And while we're combing through people's genomes for hints as to which gene is correlated with this or that trait, we are also scrambling to map the human brain and more fully understand how the complicated ballet of chemicals, cells, and structures manifest as thoughts, feelings, sensations, and actions.

Right now one of the ways we are decoding the human genome is with DNA sequencing services and databases. These companies can tell you a great deal about your ancestry, and some compare people's DNA to information they are willing to provide in surveys. This has helped locate genes for objective qualities like height, coloration, and diagnosed health problems. But if we do the same with a questionnaire about personality traits, it will be filtered through our (sometimes misaligned) sense of ourselves.

You can so easily tell the genome analysis researchers that you are five foot six, have brown hair, and developed type 2 diabetes, but how will you respond to questions about temperament, patience, positivity, stick-to-it-iveness, and so on? Will you be objective? Or will you sugarcoat the truth like so many of us do when our doctors ask how often we exercise, or the dentist asks how often we floss?

"I floss *every* day, Dr. Mouthmirror, I swear. Please don't yell at me. Okay, okay! I floss every other week or something—if that. I don't know. Just please help my mouth not be awful. And besides, it makes no sense that you're angry. Don't you get more money if I have horrible oral hygiene? Get your priorities together and take that judgmental top hat off. You look *ridiculous.*"

How will we even isolate the personality traits we are interested in? Would the questions be checkboxes where you mark all the traits you think most describe yourself? Would you be rating yourself on a patience scale between one and ten? How on earth can you boil down your entire range of behaviors into quantifiable values that you can report for use in genetic research?

I don't know about you, but some days I am the most centered, patient, high-functioning person you will ever meet, and other days I have complete mental breakdowns about banal things like laundry. (Yes, this is a real example. My husband can attest to this.)

We do not always respond to a given situation in the same way. Do we rate our personality based on an optimal environment, in which we've had enough sleep, have eaten healthy foods, have stayed hydrated, and haven't stubbed our toes? Because given an ideal situation, most people can be pretty kind, patient, and resourceful. But the same is not true of all people when they're running on two hours of sleep and half a handful of almonds in lieu of a meal, and have been stuck in traffic for hours. What's the genetic basis of that?

Eventually, we will be able to trace the genetic roots of each of your major personality traits. But people are still complicated and unpredictable. So knowing that genetically you are a patient person doesn't mean that you won't flip your lid over a roommate who for the thousandth time forgot to put the garbage cans out on trash day. Even with the genetics of distinct traits such as diseases, it is sometimes only your *likelihood* to develop them that is etched in your cells' nuclei, not a yes-or-no answer.

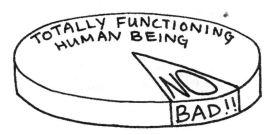

I look forward to the days when we will know more, but I hope people don't feel discouraged by the information they uncover about themselves. I feel this especially when it comes to research into the different brain patterns in women and men. All the researchers are finding are statistical averages and tendencies that don't speak for every individual. For example, research specifically looking for differences between men and women is downright counterproductive. Let's say they do find a genetic reason that men are more geared toward engineering careers (where there is still a huge gender gap). What would we do with that information? It wouldn't mean that women can't be excellent engineers—we know from experience that isn't true—but would it dissuade women who were interested, would it make sexist practices now evidence-based, and would it make people give up trying to change the status quo? I sure as hell hope not.

The genetics of personality is an interesting one. The idea of sequencing your genome and seeing a printout of your personality traits sounds so much like a horoscope to me. "According to your DNA, you are kind most of the time, you tend to be patient, and there is a 90 percent chance you like eating food and sleeping." Personality is so . . . well, personal and complicated and varied . . . that the information we uncover about ourselves will almost not even be useful.

The worst possible outcome is for people to feel hindered by their genetic likelihood of displaying less desirable personality traits. What if a future genetic test revealed that you are, in fact, predisposed to be an asshole? What do you do with that information? Would you go through life trying to go against your nature to prevent future asshole-ish episodes, the way someone genetically likely to get type 2 diabetes avoids sugar? Or would you retreat into solitude because you can't change who you are?

Do not misunderstand, though. I want to know why some people are endlessly patient and others lose their cool over a 30-second delay in the grocery line; why some people can deal with overwhelming setbacks with Pollyanna-esque positivity while those with endless resources and opportunities feel victimized by the world around them. The variation in human personality traits is staggering, and on top of that, people sometimes even change their emotional habits. Is there a genetic predisposition for adapting to your surroundings and morphing your personality to work there?

No, this entire chapter on the genetics of personality doesn't have any actual findings. But I hope that you, like me, enjoy thinking about the possibilities of this research, the ethical considerations, and the future cultural impacts. It is, quite simply, bizarre and exciting.

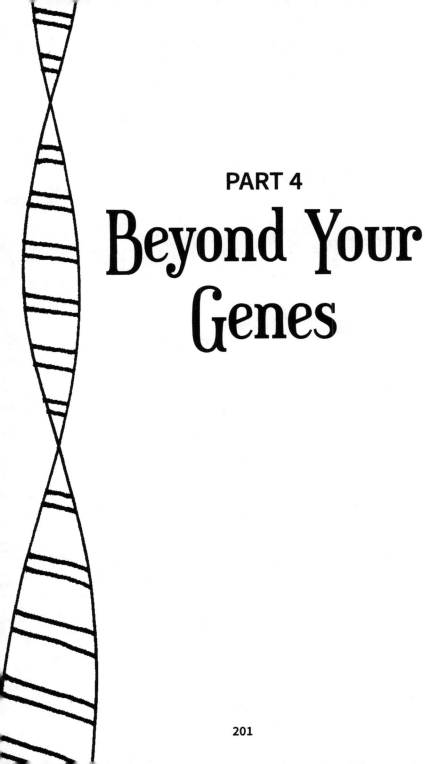

PART 4

Beyond Your Genes

Nurture Your Nature

This fight has been going on just about forever: nature or nurture. Translation: Is everything we are and will be foretold by genetic destiny, or is it all due to the effects of the environment and your experiences? As you'll often find when two sides of a fight are on completely different ends of a spectrum, the answer is somewhere in the middle.

Indeed, every time we ask this question or re-evaluate the accepted answer, it must be said that it's a combination. (Duh.)

But that's never good enough for us, is it? We need absolutes! We want answers that always work! But again, "Always and never are never true in biology." Except of course when you say it's "never true," because in that case that "never" is true. It's very contradictory, but don't worry about it.

As a society, we flip-flop on this issue a lot. At times we feel that genetics is just a starting point and that it's really how you are raised that determines the outcome. This is why when people do horrible things, sometimes we point fingers at their parents (mostly if the offender is young). But in this extreme example, it's of course not that simple. The most loving, responsible, upstanding parent in the world can't cure untreated mental illness.

Lately we've become more attuned to the limitations our genetics places on us, but that of course doesn't let parents off the hook. No matter how genetics calibrates a person for IQ, temperament, and emotional well-being, when you put them in a wildly unhealthy, abusive environment, it's going to do some real damage.

At this point you're probably saying, "Well, yeah. This isn't rocket science." But it still needs to be said! People bounce back and forth on this all the time, and let me say here and now that nature and nurture both feed off one another to make you who you are.

Nature gives us our potential and our baseline, which definitely varies widely, but it takes a nurturing environment to unwrap those qualities and bring them to life. And a craptastic environment can scramble them.

Nature, our genes, may have a big effect on who we are, but taking that into account on a grand scale can feel limiting. What if at school students weren't given tests but were asked for cheek swabs to supply DNA samples so that we could test to see their relative intelligence levels and place them in honors or remedial classes? What if we did this for sports team tryouts? Auditions for the school play?

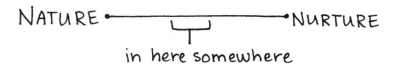

As genetic information about ourselves becomes more readily available, we as a society will have to decide how much weight we'll give it. A DNA check might be a valuable indicator for future illnesses to watch for as we age, and we are just starting to do that, but how much more information can we use without getting ridiculous?

For instance, when it comes to gun control debates, some people want an evaluation to keep firearms out of the hands of mentally unstable individuals. Sounds like a damn good idea to me. Let's say you provided some saliva and a DNA evaluation could say whether or not you have the potential for certain mental illnesses. Is that fair? Or is it unreasonable, since even a super-duper DNA scanner would only be looking at indicators, not foretelling the future with genetic certainty?

And from there things start to get really fuzzy. We obviously don't want DNA to limit us unreasonably, even though in truth it already does, sometimes subconsciously. If you come from a relatively short family, you're probably not dreaming of a life in the NBA.

Let's put it this way. DNA may predict what you *can* do, but not what you *will* do. It's just like what Dumbledore said to Harry Potter when he was disturbed by his similarities to Voldemort: "It is not our abilities that show what we truly are. It is our *choices.*"

Damn straight, Dumbledore. Damn straight.

Twinsies!

The ultimate test bed for nature vs. nurture disagreements lies in twin studies, especially when the twins were separated at birth.

I'm referring to identical twins, not just any old twins. Fraternal twins, although still twins (and therefore automatically awesome), are no more related than any two random siblings. In case you get the types of twins confused (as many people do, so don't feel bad), allow me to fully explain.

During, ahem, reproduction, after the sperms' long journey, they approached the egg(s). Some women's bodies love ovulating so very much that they release more than one egg at a time. If there are two eggs hanging around, and they both get fertilized—baboom!—you have fraternal twins. These are also called dizygotic twins, because they started out as two (hence the *di*) separate zygotes, which is what we call the first cell of a person, formed when the sperm and egg go on a very strange first date.

Fraternal, dizygotic twins are just siblings that shared the womb. I'm sure they may feel more closely bonded, spending all that fetal time together, having the same birthday (or being really lucky and being a day apart, their births having straddled midnight), and being the same age all the time. But do not be fooled! They are just sisters, or brothers, or brother and sister—whatever the case may be.

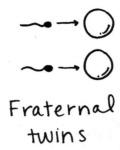

Fraternal
twins

Identical twins, however, formed from the same egg. This egg got frisky with one sperm, became a happy zygote, and started developing. But due to random chance, or a full moon or whatever, it split in two. Right down the middle. This created two separate entities with the exact same DNA. Each one developed into a person, and these people are identical twins.

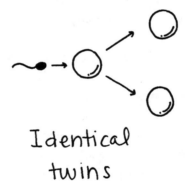

Identical
twins

There is a bit of a misconception because of certain horrible Hollywood movies (cough, *Jack and Jill*) and children's animated TV shows (cough, *Rugrats*) that there is such a thing as a boy and girl twin that are identical in every way except their sex.

No.

If a pair of twins consists of a boy and a girl, they are by definition not identical. They are definitely fraternal twins, or fraternal and sororital twins. Whatever. I don't know the rules.

Before I move on to the actual subject of this chapter, let me dispel one more television misrepresentation of twins that bothered me a tad. In *Arrested Development*, there are identical twins, George, Sr. and Oscar Bluth, that are at times only distinguishable because George is bald and Oscar is not. It's a relatively major plot point in the show. This is completely and utterly preposterous, of course, because male pattern baldness is entirely genetic, and it's impossible for one twin to be bald while his twin's head flows with hair.

I don't use words like "impossible" lightly in a book about genetics. Just about anything can happen. I'm actually now thinking of a way that a pair of identical twins could, somehow, differ in terms of the baldness gene on their X chromosomes. Maybe some sort of radiation that would in some way restore the baldness gene to the functional gene in one twin but not the other . . . no, it can't happen. It just can't. If you ever hear about identical twins that differ in terms of baldness, I want you to e-mail me. I'm serious. I'm that confident that this is completely not possible.

Back to the subject at hand: studies of twins. Twins are excellent test subjects for genetics because you can gauge how "heritable" a trait is by observing it in identical twins.

By "heritable," I mean the measure of how much genetics alone contributes to a certain trait. Discussing the heritability of something at all means that it's something complicated, like mental illness or cancer or how good you are at basketweaving—things that are a combination of genetic and environmental factors. You can gauge how heritable something is by seeing whether it occurs in both individuals in a pair of identical twins.

For instance, if identical twin sisters got breast cancer at exactly the same time, you can say that breast cancer is highly heritable. Similarly, if one twin has schizophrenia and the other twin does not, you can say that schizophrenia is not so high on the heritability scale.

Studies in this way work if you have twins who grew up together. If you have twins that were separated at birth, you have hit the nature vs. nurture research jackpot. You get to see what role the genetics of a person plays when the twins are not exposed to the same environment, as they pretty much are if they're raised together.

You also can bypass the tendency of some twins to deliberately set themselves apart and claim their own identity. Some identical twins don't much enjoy having a copy of themselves around all the time, dealing with teachers who can't tell you apart, and being so very defined by it.

One such pair of separated-at-birth twins, Jerry Levey and Mark Newman, were adopted and raised in New Jersey just miles apart. Neither of them knew they had an identical twin roaming around. They met because they were both firefighters, and a colleague of one saw the twin at a firefighters' conference. He arranged a meeting of the two, and at age 31, the brothers Levey-Newman met. The same height, the same mannerisms, the same mustache, the same bald spot. Identical twins to the nose. Even their calling to become firemen was in their DNA—or in the New Jersey water. We may never tease those two things apart.

But as much as I love this twin story and others, such as those where twins marry people with the same first names, there are also separated twin stories in which they are completely different from one another. What does this mean?

We don't know, okay?! Stop yelling at me!

But really, what it means is that genetics plays an incredibly huge role in who we are as people, but different experiences shape who we are too. Jerry and Mark both grew up in New Jersey, presumably in the same socioeconomic stratum. Identical twins that are adopted and raised separately in Spain and Antarctica could turn out very differently. That's just life, yo.

More Than Just Genes

I foreshadowed this a few times already, but prepare yourself for the genetics bombshell: Who you are isn't only about which genes you get—it also depends on whether or not those genes are turned on or off.

Let me back up for a minute. A big part of the whole Darwinian, natural selection, survival-of-the-fittest thing is that it depends only on your genes, and has nothing to do with what your parents did.

There was a scientist named Jean-Baptiste Lamarck who theorized that experiences acquired by one's parents could influence what traits they passed on to their offspring, resulting in adaptations to a specific environment.

The most referenced example is giraffes and their goofily long necks, which make drinking from a waterhole quite the awkward misadventure. Darwin explained that the giraffe's long neck was the result of random mutations that led to a population of giraffes in which the ones that happened to have the longer necks were more successful. Over time, only giraffes with ever-longer necks were able to survive, leading to the wonderful, successful, awkward species we now know and love.

Lamarck thought that a lifetime of reaching for leaves higher up on trees might have caused elongation of the neck, which would be passed down to the offspring of that individual. This example oversimplifies a fair bit, but that is the general idea.

When this spat was settled, the prevailing wisdom was that acquired traits were in no way (no how!) inherited. The genes you are given are what they are. They get put into the eggs or sperm you make, and that's that. There is no editing or refining of the DNA booklet you get from your parents. As they used to say at my preschool when handing out Popsicles of various flavors: "You get what you get."

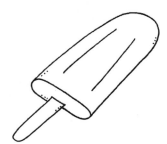

Genes are just like Popsicles.
You get what you get.

If there were a way to pass inherited traits, like your knowledge, your experience, and practice, it would be pretty handy. Think of how quickly civilization would move if each generation didn't have to spend the first eighteen years of its life learning the basics of reading, writing, and not being an asshole. If babies of PhDs were born with that knowledge, they could start researching as soon as they had the motor skills to hold a pipette and type on a keyboard. We'd have beaming technology in two generations.

But, of course, if we inherited all acquired things that would include injuries, horrible experiences, and bad habits, too. Not to mention that if you were born with your parents' knowledge, you would perhaps also be able to recall the moment of your conception, and no one wants that.

But now that we've talked about silly examples like giraffe's elongated head extenders and babies who know astrophysics, let me explain how epigenetics actually does work.

DNA can be slightly edited by things your parents do. Most possible effects are from the mother's activities, especially during pregnancy, but guys, you're by no means off the hook here. Remember, we ladies are born with all our eggs, but guys are making new sperm all the time. So if your sperm are made while you're smoking marijuana or tobacco (and anything other than that), it will affect the little tadpoles of yours.

What sorts of things can be edited by our actions? The example I mentioned earlier was giving a child asthma if its mother smokes. And it's not just smoking cigarettes that can lead to that. Breathing polluted air can also increase the risk of asthma.

Other epigenetic effects can deal with how the baby's metabolism and brain activity is calibrated. For example, a mother who undergoes extreme stress while pregnant is more likely to have a baby that grows up to have problems with anxiety and depression. For this reason, anytime I hear news about a pregnant woman who goes through some traumatic experience, my second thought (the first being "Gah, that is awful!") is "Oh, that poor child might grow up to be wired for anxiety. Poor little girl or guy."

Your weight and dietary habits can also affect your children. Being significantly overweight can affect a man's sperm and a fetus. Surprisingly, babies of overweight mommas are usually born underweight. This is likely due to a nutrient-poor diet, which can happen if you eat unhealthy crap instead of actual food.

I hope I haven't made you feel guilty about living in a smoggy city or eating potato chips before or during pregnancy. That is not what I'm trying to do here. Besides, these associations are only now being drawn, and our understanding right now is limited. What's most important is that you generally take care of yourself, whatever that means for you.

What Fruit Flies and Others Tell Us

The most famous genetics researcher, our buddy Gregor Mendel, used pea plants to discover the fundamentals of heredity. Today, geneticists aren't tending vast gardens of pea plants and brushing pollen onto stigmas. They're inside, tending to many generations of living model organisms to understand more about genes, the RNA and protein they make, and how they affect the entire organism. Among the living things they study: fruit flies, rodents, zebra fish, yeast, and bacteria.

Fruit Flies

The term for someone who spends a lot of time at the lab bench is "lab rat," but it would be more accurate if it was "lab mouse," since they are used with such frequency. But when it comes to genetic research, you may be surprised that one of the most useful organisms is not the mouse or a rat, or even a mammal of any sort. It's a fruit fly.

Yes, a fruit fly. Those little jerks that show up when I hold onto overripe bananas and tell myself I'll use them to make banana bread but never do. And here's a pro tip: if they take over your kitchen, put apple cider vinegar (or red wine) in a bowl, cover it with saran wrap, and punch holes in it with a fork. The fruit flies will be helpless against the luring of the vinegar, and will enter through the little holes in the saran wrap and won't be able to find their way back out. After a few days, you'll have a bowl full of them floating in the vinegar. Done!

But if you work in a genetics lab, odds are you won't want to murder all the fruit flies; you'll want to breed them and count them. All day long.

Why would we spend so much time doing genetic studies on fruit flies, of all things? I know it might sound strange, since they don't seem closely related to us. But stop right there. Just because they're not our close relatives, evolutionarily speaking, doesn't mean that we can't learn a great deal about genetics from them. It's not all about us humans.

There are a few reasons fruit flies make such great genetics specimens:

» They're small and easy to keep in a lab.
» Their generation time is just a few weeks, and females can lay about 100 eggs at a time.
» It is relatively easy to tell the difference between boys and girls, so you can separate them out to control breeding.

» They only have four chromosomes, so that simplifies things when mapping their whole genome.

» They have a few easily distinguishable traits that you can work with in the lab, like white eyes vs. red eyes. If you want to study a different gene, you can attach it to the gene for white eyes, and in that way know at first glance which flies have the gene you're looking to test, as if the fly is holding up a little tiny picket sign that says, "Pick me!"

Back when I mentioned baldness and colorblindness, I said that these traits are sex-linked, meaning that they are not inherited completely independently, but instead their incidence is related to the sex of the organism. It was studies with fruit flies that first showed this. They're so helpful.

Mice

Among the mammalian group, it's mice that are the workhorses of genetic laboratory studies. If horses were the workhorses, labs would have to be pretty large, and they'd need a team of poop scoopers.

Mice are useful model organisms for many of the same reasons that fruit flies are. They are small, easy to care for, easy to handle (although they probably nip you sometimes), reproduce often, and mature very quickly. They're also cute.

So what are we doing with these little guys? One of the common ways to study genes is to make knockout mice.

Far from prizefighter, boxing mice (as the name might suggest), knockout mice are engineered to be lacking a specific gene. Often the easiest way to learn the function of a gene is to see what happens when an organism doesn't have it. Thousands of genes have been tested this way with mice so that we could pin down their exact purpose, simply by seeing how the mouse is impacted without that gene.

Mice are also good test subjects because they are so similar to humans, which means they get many of the same diseases that we do. You can use a mouse to model a human genetic disease by changing the mouse gene that corresponds to the responsible human gene. This is true for cancer, heart disease, hypertension, diabetes, obesity, osteoporosis, and deafness, just to name a few.

I know it sounds sad that researchers are giving mice cancer or making them overweight on purpose, but it's not as grim as it sounds. While I have never conducted research with animals, I do know people who do. The care for these animals is as well thought out and meticulous as anything else scientists do. Even though these little furry fellows have diabetes and hypertension, they have Cadillac health insurance.

Zebra Fish

Zebra fish are helpful in the lab in a way that flies and mice aren't— early in development they're transparent. That means that if you are looking at the way a gene (or the lack of a gene) is affecting the growth of a particular structure, you can actually watch it happening under a microscope in just a few days. And since our genomes are 70 percent the same, there are lots of conditions or analogs for those conditions that we can test with them.

Scientists have taken advantage of the transparency of zebra fish by using fluorescent markers—making some parts of them glow.

When studying how zebra fish fight off infection from bacteria, a researcher can introduce fluorescent bacteria and watch in real time which zebra fish get glowing infections and which don't.

You want to see if a certain protein or gene is being expressed? Add a fluorescent marker to it and see where it shows up! You can even buy these luminescent zebra fish in pet stores now.

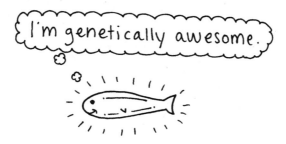

Yeast

On to our bread-and-beer-making friends, the yeast. This species has long been a very helpful lab mate. For one thing, yeast are incredibly easy to grow. You may have done this if you've made bread from scratch, or brewed your own beer. Give yeast a warm water bath, and they'll be, as my grandfather used to say, "as happy as a lark."

Yeast are single-celled, but they are eukaryotic, which means that, like us, they have a nucleus and organelles, so even though they look nothing like us, we still have a great deal in common, biologically speaking.

Yeast have been especially helpful in modeling how mishaps in the cell division timeline can lead to cancer, which happens when cells divide uncontrollably and ruin everything. Because yeast cells are so similar to ours structurally and in their regulation of cell division, we can learn exactly how things can go wrong in this area in our own bodies.

Bacteria

And lastly, our bacterial friends. They too are easy to keep in a lab space. Put them on a petri dish filled with nutritious agar jelly, crank up the thermostat, and they will go to town, growing into adorable little colonies.

A bacterium is different from the rest of the organisms mentioned in this chapter because it is a prokaryote: it does not have a nucleus. Its DNA is not housed in a special nuclear house—it's out in the cell. It also doesn't have the vast library of noncoding DNA like we do. Bacteria only keep the DNA they use.

But one thing that bacteria happily do is use any DNA that you give them. Introduce a gene into a bacterium, and watch it go. Scientists did this with the gene for insulin and sat back as the bacteria produced the protein like there was no tomorrow. That's one of the ways that we commercially produce insulin for diabetic folks that need it so badly. Thanks, bacteria!

Bacteria also helped us see the basics of how gene regulation works—how genes can be turned on and off depending on whether you need the protein it codes for. The textbook example is the *lac* operon—a segment of DNA that controls the expression of a nearby gene. *Lac* stands for lactose—a sugar found in milk.

I'll give you the short version of how these gene switches work. The *lac* operon controls a gene that codes for enzymes that break down lactose. When there is no lactose around, the gene is turned off because a "repressor molecule" is attached to the DNA, preventing it from being expressed.

But when there is lactose in the cell, it actually binds to the repressor molecule, which makes it fall off the DNA. When this happens, the gene for breaking down the lactose gets expressed. Now the cell makes proteins that metabolize lactose.

And here's where it gets really smart: once all the lactose has been eaten up, there isn't enough to keep the repressor molecule busy, so it reattaches to the gene and turns it off again. Because all the lactose is gone, there's no need to keep making proteins that eat it.

This is a great system. It works just like a thermostat. If you set your thermostat to 70 degrees (or 21 degrees if you're in a Celsius part of the world), the heater will turn on when it drops below 70 (or 21), and it will turn off if the temperature goes above 70 (or 21). It has a mechanism in place to maintain a happy medium. And bacteria do too. They're pretty clever that way.

You, the Test Subject

The other valuable test bed for genetics is ourselves. While we don't actively test on people, as that would mean giving them genetic diseases on purpose (although plenty of egregious ethical lapses have happened in the past regarding testing on humans), we can learn a lot from the random happenings of our genetic lives. Modern technology gives us new ways to peer inside our cells, giving us insight into the inner workings of our DNA, leading to a deeper understanding of our evolutionary past as well as medical breakthroughs that can save lives.

And you (yes, you!) can help with this if you're interested. I'll talk more about this new area of genetic research later in the book.

Comparative Genomics

We got our DNA from our parents. They got it from their parents, who got it from their parents, who got it from their parents . . . I could go on like this for a few pages, but I'll spare you. Let's just say that it goes on and on like that until we get to the first little single-celled life blob that blooped into existence 3.8 billion years ago.

It should come as no surprise then that DNA contains information about our evolutionary past, and the more similar two organisms' DNA is, the more evolutionarily related they are, such as the oft-cited degree of similarity between ourselves, chimpanzees, and bonobos. We great apes share about 98 percent of our DNA. That other 2 percent is what codes for genes that give us bigger brains, speech, and the disinclination (in most circumstances) to throw our feces.

Comparative genomics—looking at the similarities between different species' DNA—has given us a new way to peer into our evolutionary past. We already could tell the relative relationships of different species on the tree of life (the vast network of evolutionary connections between all life on earth) based on physical traits of those creatures. But DNA has shed more light on those relationships, reaffirming our evolutionary closeness to other organisms. And not just our close cousins in the great apes—chimpanzees, bonobos, gorillas, and orangutans—but in the entire family tree including animals, plants, fungi, protists, bacteria, and archaea. We're all related.

The differences in DNA are due to mutations that build up over time. Overall, mutations are random and occur at a constant rate, so when we compare two species' DNA, more differences mean more time separates them evolutionarily. This constant changing of DNA over time that leads to new species is called the molecular clock. It's a very small clock that measures some very big things.

When we talk about genetic similarity, there are a few things we can look at: specific genes, spaces between genes, or the entire genome. These comparisons come out very differently.

Sometimes the best way to get a glimpse of two species' similarity is in spaces between genes, or the not-so-junk DNA I mentioned earlier in the book. In these portions of DNA, there is nothing to dissuade mutations from building up. If a mutation happens in an important gene, like the one that makes the ribosome protein factories of your cells, it's going to be bad news for the poor little creature that got that mutation. And that means the little guy probably won't have babies, which means that mutation is lost to the ethers, and for those of us looking for mutations right now, that's not helpful.

Basically, important genes are subject to selection and therefore have to work right, or they don't continue. Noncoding DNA is often not subject to this pressure to perform, so it accumulates mutation after mutation at the normal rate, with nothing to stop it. It comes in handy for scientists, though, because that gives the most accurate reading of when two species split on the evolutionary tree.

What else can our DNA tell us?

We can see what diseases our ancestors contended with. Because viruses' favorite pastime is injecting DNA into our cells, some of it gets retained and passed down to future generations. In that way, we can actually trace when big disease epidemics happened—thousands or millions of years ago—based on which species have the viral remnants.

One type of viral marker is 100 million years old, and is found across thirty-eight different types of mammals—from humans to mice, elephants, and dolphins. That means that it was a common ancestor of all of us that was infected with it. That grandmother that you and I and Flipper share must have had a wicked cold. A different virus infected a more recent common ancestor of ours, so its mark is only left on us and other primates like chimpanzees and bonobos.

Our connection to our ancestors of other species is not an abstract, hippie-dippie dream. It is real, it is physical, and it is inside you. It makes me want to run naked through a meadow and make moose noises. I may be alone in that impulse, though.

Who Owns Your Genes?

Scientific research is in many ways based on capitalistic practices. Researchers compete for funding, race to be the first to publish results, and in every other way strive to be the best. If you're discovering new things, part of this process involves patents. Most of the time, scientists want to financially benefit from the work they've done. There's nothing wrong with that, of course; it's how our entire economy is set up. But patenting a new kind of light bulb is one thing—doing genetic research and patenting specific portions of DNA is a whole other can of genetically modified worms.

Up until June 2013, researchers could patent genes they discovered or altered. This has come into play in areas like genetically modified crops such as roundup-ready corn. Monsanto owns the patent of that species' genes because its researchers made the changes to the genetics of the corn plant, so they own the rights to all those genes, wherever they may be.

This gets particularly thorny (hehe) with plants because we're not talking about a piece of machinery or some newfangled gadget; we're talking about a living thing that under normal circumstances spreads across the land when seeds are blown by wind to neighboring areas. Having roundup-ready corn found on your property that you didn't pay for is a very serious offense. But is it a farmer's fault if a seed from a nearby farmer was blown onto his land? This is complicated, stressful stuff, and we don't have it all figured out yet.

Our society and our government are still working the kinks out of this relatively new area, where researchers can create new genes and subspecies of modified organisms and want to, of course, protect their work and investment. But we also have to remember that these are living things, and to quote *Jurassic Park* once more, "Life cannot be contained."

Enter the penguins—I mean, justices.

Yes, the Supreme Court of the United States passed a decision in June 2013 that naturally occurring genes *cannot* be patented. Score one for those old farts! Excellent work, justice league.

If we had continued down a path of patenting every gene we sequence, it wouldn't have been pretty. It would have been downright disastrous for research, health care, and squirrels. (Just kidding about the squirrels.)

But seriously, imagine that a genetic disorder runs in your family, and early detection is important for survival. Now imagine that the gene that is responsible for this condition has been patented by Dr. Soandsoface. The genetic test costs $10,000 because part of the testing cost goes toward satisfying the fees associated with the patent. Can you afford to drop $10,000 on this test? That's obviously not a reassuring, heartwarming medical situation.

Without patents on genes, no one owns the rights to a specific portion of DNA. When the gene that causes your hypothetical familial disorder is discovered by research scientists, it is published in journals, and now every doctor in the nation knows how to test you for it. The test is covered by your insurance with a nominal copay. That plan sounds better to me.

This situation is exactly what was happening and what will likely happen with screening for breast cancer genes. Prior to this Supreme Court ruling, those genes were patented; and the tests for them, expensive. Now that you can't patent a breast cancer gene, tests screening for them will be open sourced and, hopefully, far more affordable than they are now.

Genes can still be patented, but only if researchers create something novel. This sounds more in line with my fourth-grade understanding of patents. You take out a patent if you spend time and money to invent something completely new, not if you find it in the woods: just as no one can patent a reindeer because they stumbled upon it, no one owns the rights to your genes.

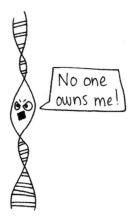

The related issue here is this: Who owns your *cells?*

The most famous case of cell ownership is Henrietta Lacks. She lived in Virginia from 1920 to 1951. While she was in the hospital with cervical cancer (which eventually killed her), a cell sample was taken from the tumor. These cells were grown in a lab and were remarkably willing to keep growing and dividing.

This cell line was called HeLa after the original owner of these cells, but no one asked her permission for them, and her children and grandchildren for the longest time didn't know that her cells lived on in petri dishes, providing a valuable means for experiments that led to some of the most brilliant medical breakthroughs in history, making some people incredibly wealthy while Henrietta Lack's surviving family struggled in poverty.

Consent, permission, and comprehension of what it means to donate your cells to scientific study raise questions that our species has never before had to consider. Do you cease to own your cells the moment they leave your body? If a researcher uses a vial of your blood to unlock a medical breakthrough, are you in any way responsible (and owed money)?

Right now the answer is no, for several reasons. For one thing, the identity of tissue samples is usually kept confidential. Naming a line of cells after the person who made them, like in Henrietta

Lacks's case, is not the norm. Although a line of cells called KaMc does have a nice ring to it . . .

Also, setting a precedent where every person who donates tissues that can be used in medical research is owed money down the line presents a significant hindrance to the scientific enterprise. Not to mention a lot of paperwork.

But again, I'm just talking about people who willingly give tissue samples or agree that samples taken during medical procedures can be used in research. The situation is very different if cells are taken without your knowledge. That's just creepy and horribly uncool.

This is a tricky subject, and one with pending lawsuits involving angry patients and hard-working scientists. But as of now, legally, once cells leave your body, you no longer own them and cannot dictate what happens to them or what research they are used for. Many hospitals and doctors provide a consent form that you sign to authorize them to keep your cells and possibly use them for research. But they actually aren't legally required to ask your permission. If you willingly hand over cells—give a blood sample, pee in a cup, etc.—they can do whatever they want with them.

You *do* own the reproductive cells outside of your body, though. If you have eggs or sperm in a frozen bank, those are very much yours. All of them! Happy sperm and eggs!

Get Your Sequence!

The Human Genome Project lasted thirteen years. That's how long it took to determine the sequence of the more than 3 billion bases in a human being's DNA. For the longest time I had a basic misunderstanding about this Project—I thought they were reading the sequence of a single person's DNA. My high school self and my deep understanding of DNA and the scientific enterprise scoffed at this: *Pffft. They're not unlocking the human genome; they're just mapping one single person's genome. What good is that? I bet his name is Dave and he makes horrible pancakes.*

Turns out, I was very much mistaken. Imagine that. A teenager who thought she had all the answers was wrong! Shocking.

They weren't sequencing Dave Whatshisface's DNA—they were sequencing several different mixed genomes from anonymous donors' blood and sperm.

No one knows who these donors were—not even the donors themselves. It's all very secretive.

But these days, you don't have to have thirteen years and billions of dollars to find out what's going on in your molecular world. It's a very exciting time. You can peer into your DNA for as little as $100 via services like 23andMe and National Geographic's Genographic Project.

The aim of services like 23andMe is to sequence your DNA and find possible relatives among their database of participants (if you choose to enroll in that part of the process), and they also can tell you whether you have some of the currently known cancer genes and other diseases, as well as things you already know about yourself, like your genes for hair color, eye color, and whether cilantro tastes like soap to you.

The Genographic Project is somewhat different in scope. It is more focused on ancestries and our evolutionary origins. If you participate in this project, you get a reading of your DNA that gives you the percentages that you are related to specific genetic groups of people. There are nine ancestral components they consider: Mediterranean, Northern European, Southwest Asian, Southeast Asian, Northeast Asian, Native American, Southern African, sub-Saharan African, and Oceanian.

They can also tell you how much of your DNA is from Neanderthals or Denisovans, extinct human relatives that interbred with the earliest *Homo sapiens.*

The ladies love
that thick brow.

The extra benefit of most DNA assessment services is that it lets you participate in citizen science projects, where the contributions of many individuals help scientists conduct their research. By providing DNA from simple cheek swabs and sending them to big projects like this, scientists can use the information to conduct studies and discover new phenomena. It's all-around pretty fantastic times we live in. It's a big science party, and the whole world is invited.

The Future of Genetics

The future is now (as Jim Carrey so eloquently put it in *The Cable Guy*). The dawn of a new type of medical treatment is at hand: gene therapy. Except it has nothing to do with talking-it-out, blame-it-on-your-father, actually-just-take-this-pill kind of therapy. It's therapy for your genes, as in changing them. Yes, raise those eyebrows and widen those eyes.

Changing your genes? But Katie, you said that couldn't happen. You said crossing the streams was bad (*Ghostbusters*). Yes, I probably said something along those lines—that your genes are yours no matter how much you may hate them, that they cannot be edited, that you get what you get.

I lied.

Doctors are testing new ways to replace faulty genes we have with shiny, functioning, new ones. It's not easy, and it will take time to work out the kinks, but the possibilities are mind-blowing. Think of it—the end of genetic disease. The end of blindness. The end of toe hair. My god, it will be beautiful.

How will such a feat be accomplished? With viruses.

Viruses specialize in sneaking foreign nucleic acids (DNA and RNA) into cells. It's what they do. It's pretty much all they do, those little jerks.

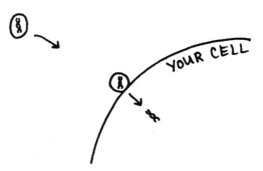

YOUR CELL

But we can harness this power and use it for good. If you design a virus and give it a good gene, and deliver it to the area where that new gene needs to be incorporated into cells (such as in the retina if you're reversing blindness), then presto. You have turned the virus into a gene deliveryman who doesn't even ask for a tip.

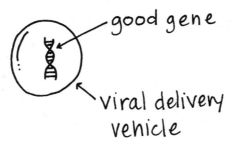

good gene

viral delivery vehicle

And in case you're thinking of this—yes, I know that this is how zombie movies begin. Headlines read "miracle cure for all diseases found" and the next day everyone's eating each other and the walking dead are shambling down deserted city streets. But come on. We're not giving these viruses genes that can reanimate dead tissue or rot people's brains. Those don't exist, so the possibilities of zombie apocalypses are minimal.

The sci-fi movie you *should* focus your paranoia on is *Gattaca*, the vision of our all-knowing genetic future. In case you haven't seen it (as I'm not quite sure if it's still a pop culture prerequisite), in the future people don't have babies by playing the genetic roulette and entrusting the baby's future to whatever random egg is lying around and whichever sperm gets there first. No, no, no. People go to genetic counselors and choose which of their genes they want the baby to inherit and have the opportunity to correct any genetic "mistakes" while they're at it.

As a result, everyone is gorgeous and smart, and no one needs glasses. I'm not just saying that because my myopic self is jealous—it's a major plot point because Ethan Hawke, the product of "natural" reproduction, has bad vision that needs correction. This fact makes him stick out as someone who doesn't belong, for in this world of genetically perfect people, those with any defect are societal outcasts. Don't let anyone know you're wearing contacts, Ethan! Be careful! Nooooooo!

But a future where a newborn baby's DNA is sequenced and a birth receipt prints out that lists all the diseases he'll get, along with the date and cause of death, is not likely any time soon, or really at all. Like I have said a dozen or so times in this book: your genes are not everything. What happens to you in life depends on what you do. Where you live, what you eat, what books you read—this stuff matters. And even in a future where we know every secret kept inside the DNA treasure chest, you can't be certain about the age of someone's death. They could always be hit by a bus.

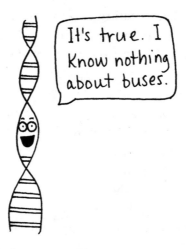

What will likely happen is that we can read our DNA (although we may not do it in the delivery room), and get a better handle on how to take care of ourselves. Doctors could use your genetic dossier to better inform decisions about your medical care, and individuals could get a glimpse into their hidden selves.

We're seeing the beginning of this in those who test for breast cancer and opt for a double mastectomy. This is perhaps the most intense form of genetically informed decision-making. And it exposes the underlying human problem: we're problem-solvers, we *Homo sapiens*, so having a future issue genetically foretold to us is going to make us want to take action. I'm prone to breast cancer? Lop those suckers off. I'm in danger of developing diabetes? I don't want any birthday cake, thank you. I'm going to be completely bald by the time I'm 35? How much is Rogaine?

But where does the reasonable prevention end and the paranoia and overreaction begin? We're going to have to figure that out. But I'll tell you one thing: people need to understand the fundamentals of genetics to make those decisions. It's going to require a broad yet nuanced comprehension of how our bodies work, how our genes lead to traits, and why DNA is so goddamn lazy.

As genetic and medical research advance and intermingle, our decisions about our health and that of our family will need to be informed by our understanding of the basic principles I have described in this book. For like most technology, genetic research can be a source of hope, or it can be our downfall, depending on how we use it.

No pressure or anything.

Sources and Further Reading

I'm not a great team member for trivia night at a bar, as my general knowledge base does not have great breadth, but I do remember my random science factoids for the most part. For this book, I made an outline of my accumulated knowledge about genetics and fact-checked myself. That's how I discovered that earlobes are not a single-gene Mendelian trait, for example. On that topic, I came across the website of Biological Sciences Professor John McDonald from the University of Delaware about the myths of human genetics. He and I exchanged a few e-mails on the subject that better informed my coverage here.

My online sources for this book include the following:

Online Mendelian Inheritance in Man: An Online Catalog of Human Genes and Genetic Disorders

www.omim.org

National Center for Biotechnology Information

www.ncbi.nlm.nih.gov

Human Genome Project Archive

http://genomics.energy.gov

23andMe Genetic Testing
www.23andme.com

Genetics Home Reference
http://ghr.nlm.nih.gov

Human Genome Organisation
www.hugo-international.org

Human Gene Nomenclature Committee
www.genenames.org

Genographic Project
https://genographic.nationalgeographic.com

Family Tree DNA
www.familytreedna.com

Scitable by Nature Education
www.nature.com/scitable

Books

The main text I used was:

Emery's Elements of Medical Genetics, 14th edition by Peter Turnpenny and Sian Ellard

For further reading about the human body and our connection to the rest of the tree of life, I highly recommend the following books that I have personally very much enjoyed, most of which also informed this book in one way or another.

Zoobiquity by Barbara Natterson-Horowitz and Kathryn Bowers

Sex Sleep Eat Drink Dream: A Day in the Life of Your Body by Jennifer Ackerman

Your Inner Fish by Neil Shubin

Riddled with Life by Marlene Zuk

The Immortal Life of Henrietta Lacks by Rebecca Skloot

Sex at Dawn by Christopher Ryan and Cacilda Jethá

Glossary

Adenine: One of the bases in DNA that make up the rungs of the ladder. It pairs with thymine.

Allele: A form of a gene. The gene for flower color, for instance, may have a pink allele and a purple allele.

Amino acid: The building blocks of proteins.

Base: The part of DNA that forms the rungs of the double helix ladder.

Chromosome: A structure made of tightly wound DNA.

Codon: A set of three bases in RNA that translates to one amino acid when proteins are being made.

Cytosine: One of the bases in DNA that make up the rungs of the ladder. It pairs with guanine.

Deoxyribose: A pentagon-shaped sugar that is found in the backbone of DNA.

DNA: An acronym for deoxyribonucleic acid, which is really boring stuff that is found in all of your cells and contains all the information for making you who you are. You know. No big deal.

Dominant: What we call a form of a gene when it gets expressed and can cover up another form of the gene (since we all have two copies of each of our genes).

Gene: A section of DNA that contains instructions for making RNA. That RNA may even be used to make a specific protein.

Genetic code: The key our cells use to translate 3-base sections of RNA into protein.

Guanine: One of the bases in DNA that make up the rungs of the ladder. It pairs with cytosine.

Mutation: Any change in DNA. It could be very small, just one letter, or it could be a missing or extra segment.

Nucleotide: The basic building block of DNA. It is made of a sugar, a phosphate, and a base.

Phosphate: A chemical compound that makes up part of the backbone of DNA. It is a phosphorous atom surrounded by four oxygen atoms.

Protein: A chain of amino acids that twists and folds into an absolutely insane shape. They do just about everything of importance in your body.

Recessive: An allele that can get covered up by a dominant one. A recessive trait is only expressed if you receive two copies of it.

Replication: The process of DNA making a copy of itself.

Ribose: A pentagon-shaped sugar that is found in the backbone of RNA.

Ribosome: A small protein factory that is found in the main part of a cell. It is made of two RNA subunits.

RNA: An acronym for ribonucleic acid, which is usually a single-stranded string of bases that is made from a sequence of DNA. Some RNAs do jobs in the cell, and others are used to make proteins.

Stop codon: A 3-base "word" in the RNA sequence that tells a ribosome to stop building the protein.

Thymine: One of the bases in DNA that make up the rungs of the ladder. It pairs with adenine.

Transcription: The process of a strand of RNA being produced according to the sequence of a section of DNA.

Translation: The process of a ribosome reading a sequence of RNA and using it to build a protein.

Uracil: A base (like adenine, thymine, cytosine, and guanine) that is only found in RNA, not in DNA. It takes the place of thymines in RNA, so it pairs with adenines.

Virus: A piece of nucleic acid (sometimes DNA and sometimes RNA) surrounded by a protein shell.

Index

About the Author

I grew up in Reno, Nevada (Biggest Little City! Woop woop!), in a big family—I'm the second of four children. I went to the University of Southern California first as a chemistry major, and then I switched to biology because I'm crazy like that. I also got a master's in science teaching because I wanted to be a pastry chef. Just kidding. I wanted to be a

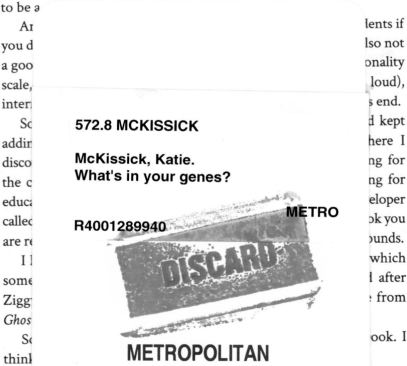